Cracking the High School Math Competitions

Covering AMC 10 & 12, ARML and ZIML

Kevin Wang, Ph.D.
Kelly Ren
John Lensmire

PUBLISHED BY ARETEEM INSTITUTE
WWW.ARETEEM.ORG

ISBN: 1-944863-00-1
ISBN-13: 978-1-944863-00-5

First printing, January 2016.

Contents

1 Introduction 7

1.1 About Areteem Institute 7

1.2 Interest is Fundamental in Learning 8

1.3 Mathematical Competitions 8

2 Number Theory 11

2.1 Number Theory Fundamentals 11

2.1.1 Warm-up Questions 11

2.1.2 Place Values and Number Bases 14

2.1.3 Divisibility 17

2.1.4 Divisibility - continued 20

2.2 Modular Arithmetic and Further Topics 23

2.2.1 Modular Arithmetic 23

2.2.2 Modular Arithmetic - continued 27

2.2.3 Pythagorean Triples 33

2.2.4 More Challenge Problems 39

2.3 Number Theory Techniques **45**

2.3.1 Problem Solving Techniques in Number Theory 45

2.3.2 Practice Problems 46

3   Algebra 59

3.1 Factoring Polynomials **59**

3.1.1 Find Common Factors 59

3.1.2 Use formulas 61

3.1.3 The UNFOIL Method 63

3.1.4 Factor by Grouping 65

3.1.5 Split and add terms 66

3.2 Solving Equations **68**

3.2.1 Quadratic Equations: Discriminant and Vieta's Theorem 68

3.2.2 Equations with Fractions 77

3.2.3 Equations with Absolute Values 78

3.2.4 Equations with Radicals 79

3.3 Factoring Polynomials Continued **86**

3.3.1 Change of variables 86

3.3.2 Finding roots 88

3.3.3 Method of undetermined coefficients 89

3.4 Polynomials **93**

3.4.1 Definitions related to polynomials 93

3.4.2 Facts about polynomials 94

3.4.3 Useful Theorems 97

4   Geometry 105

4.1 Angles **105**

4.2 Area Methods **115**

4.2.1 Fundamentals 115

4.2.2 Two Theorems Using Areas 121

4.3 Circles **128**

4.3.1 Fundamentals 128

4.3.2 Arcs and Angles 132

4.3.3 Power of a Point 136

4.4 Solid Geometry **139**

4.5 Trigonometry **149**

5   Combinatorics 163

5.1   Combinatorics Fundamentals **163**

5.1.1   Counting Basics 163

5.1.2   Advanced Counting 169

5.1.3   Binomial Coefficient Identities 175

5.2   Further Practice on Counting **179**

5.3   Probability Concepts **186**

5.3.1   Definitions 186

5.3.2   Classic Model and Geometric Model 187

5.3.3   Principle of Inclusion-Exclusion 189

5.3.4   Independence 191

5.3.5   Conditional Probability 193

5.3.6   Total Probability and The Bayes Formula 194

5.4   Distribution and Expected Values **206**

Answer Keys to Problems 215

1. Introduction

This book contains the training materials of the Math Challenge courses from Areteem Institute.

1.1 About Areteem Institute

Areteem Institute is an educational institution that develops and provides in-depth and advanced math and science programs for K-14 (High School, Middle School, Elementary School, and 2-year college) students and teachers. Areteem programs are accredited supplementary programs by WASC (Western Association of Schools and Colleges). The curricula meet and exceed the Common Core State Standards. Students may attend the Areteem Institute part time or fulltime through these options:

- Live and Real-time Face-to-Face Online classes;
- Self-paced classes by watching the recordings of the live classes;
- Summer and Winter Camps.

The Areteem courses are designed and developed by educational experts and industry professionals to bring real world applications into the STEM education. The programs are ideal for students who wish to perform well in math competitions (AMC, AIME, USAMO, IMO, ARML, MATHCOUNTS, Math League, Math Olympiad, ZIML, etc.), Science Fairs (County Science Fairs, State Science Fairs, national programs like Intel Science and Engineering Fair, etc.) and Science Olympiad, or purely want to enrich their academic lives by taking more challenges and developing outstanding analytical, logical thinking and creative problem solving skills. AP and SAT/ACT courses are available as well. Classes are usually offered in groups. Private coaching is available to students who take the group lessons and are recommended by the instructors.

Since 2004, Areteem Institute has been teaching with the methodology that is highly promoted by the new Common Core State Standards: stressing the conceptual level understanding of the math concepts, problem solving techniques, and solving problems in the real world applications. With the guidance from experienced professors with infusing passion and enthusiasm, students are encouraged to explore deeper by identifying an interesting problem, researching it, analyzing it, and using critical thinking approach to come up with multiple solutions.

1.2 Interest is Fundamental in Learning

The most important aspect of teaching mathematics is to protect the interest level in the students. For a teacher, the students' interest in learning is most precious in the world. Interest is key that opens the gate to mathematical treasures. Study with high interest is the exciting exploration, and the process of finding solutions is a great adventure. In others' eyes, it is sweat and hard work, but for the explorer, it means enjoyment and fulfillment.

Achievements in math competitions, such as AMC 10 and 12, can be big boosts in the students' interest. However, it is more essential to maintain continuing interest after the competitions. Every student is good at math to begin with, but everyone responds differently to the materials and the way of teaching. A good teacher can find ways to bring out the mathematical potential from each student according to their own special circumstances.

Many instructors give big, long lectures without consideration of how the students have learned. In those classes, the students are observers, not active participants. In case the student loses interest in one class session or one topic, it might affect the entire subject. For effective teaching, the instructor first has to understand the student, to have a complete picture of the student's mathematical knowledge and academic skills, and guide the student to discover the new knowledge on his or her own. Many ancient educators, including Confucius and Aristotle, recognized the human variabilities in the education process, and adapted their teachings to the individual needs and capabilities of the students. Individualized instruction is vital in the success of mathematical education.

1.3 Mathematical Competitions

Areteem Institute has a long history of supporting and promoting mathematical competitions in the United States. One of the authors, Kevin Wang, has been problem writer, reviewer, and grader for American Mathematical Competition (AMC) for years.

AMC is the series of the math competitions for selection of national mathematical olympiad teams. It also serves the purpose of popularize mathematics and problem solving in middle schools and high schools in the United States. The format and structure of the different levels of the competitions are shown in the image next page.

The math competitions for middle and high school students generally do not involve college mathematics such as calculus and linear algebra. There are four main topics covered in the competitions: Number Theory, Algebra, Geometry, and Combinatorics. The problems in the math competitions are usually challenging problems for which conventional methods are not sufficient, and students are required to use more creative ways to combine the methods they have learned to solve these problems.

This book covers these topics in the next chapters, along with fundamental concepts required and problem solving strategies useful for solving problems in the math competitions.

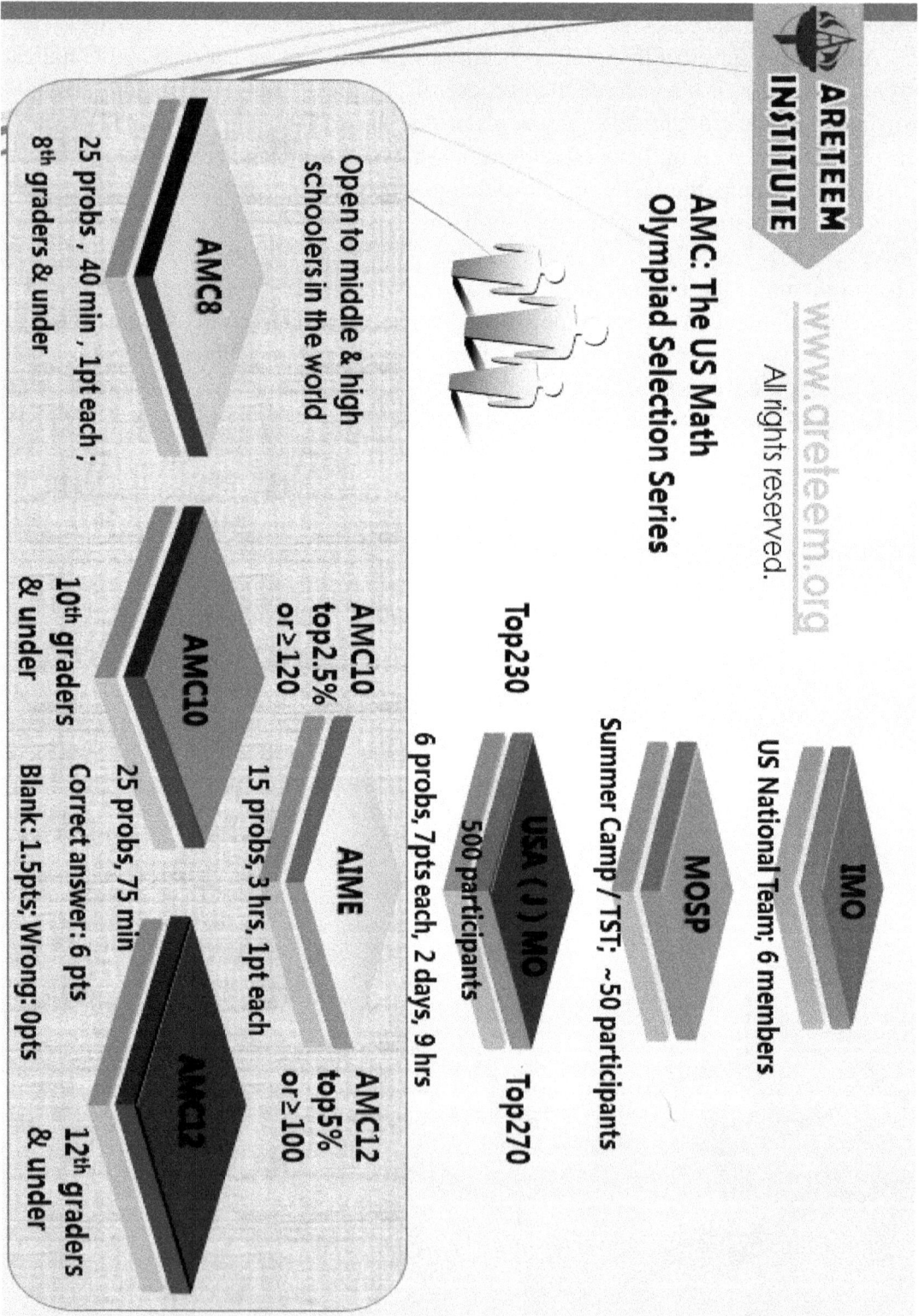

ARETEEM INSTITUTE

AMC: The US Math Olympiad Selection Series

Open to middle & high schoolers in the world

AMC8

25 probs, 40 min, 1pt each; 8th graders & under

AMC10

AMC10 top2.5% or ≥120

25 probs, 75 min

10th graders & under

Correct answer: 6 pts

Blank: 1.5pts; Wrong: 0pts

AMC12

AMC12 top5% or ≥100

12th graders & under

AIME

15 probs, 3 hrs, 1pt each

USA (J) MO

6 probs, 7pts each, 2 days, 9 hrs

500 participants

Top230

Top270

MOSP

Summer Camp / TST; ~50 participants

IMO

US National Team; 6 members

2. Number Theory

2.1 Number Theory Fundamentals

2.1.1 Warm-up Questions

Problem 2.1 The equation

$$62 - 63 = 1$$

is obviously false. Can you move only one digit to make the resulting equation true?

Problem 2.2 Can you find ...

(a) A multiple of 1350 that is a perfect cube? (You have *one second* to give an answer)

(b) The smallest positive multiple of 1350 that is a perfect cube?

(c) All positive multiples of 1350 that are perfect cubes?

Problem 2.3 One hundred light bulbs were labeled 1,2,...,100, each controlled by one switch. At the beginning all light bulbs were turned off. One hundred students, who had nothing better to do, started flipping switches. Suppose the first student flipped every switch; the 2nd student flipped the switches of light bulbs labeled with even numbers; the 3rd student flipped the switches labeled with multiples of 3; the 4th student flipped those labeled with multiples of 4, and so on. The last student only flipped the switch labeled 100. Which light bulbs are on at the end?

Problem 2.4 Diophantus was one of the last great Greek mathematicians; he developed his own algebraic notation and is sometimes called "the father of algebra." This riddle about Diophantus' age when he died was carved on his tomb:

> God vouchsafed that he should be a boy for the sixth part of his life; when a twelfth was added, his cheeks acquired a beard; He kindled for him the light of marriage after a seventh, and in the fifth year after his marriage He granted him a son. Alas! late-begotten and miserable child, when he had reached the measure of half his father's life, the chill grave took him. After consoling his grief by this science of numbers for four years, he reached the

end of his life.

How long did Diophantus live? (Can you do it without algebra? Can you do it in three seconds?)

2.1.2 Place Values and Number Bases

Place Values. The value of a digit depends on its place, or position, in a number. For example,

$$465 = 4 \times 100 + 6 \times 10 + 5$$
$$2013 = 2 \times 1000 + 1 \times 10 + 3$$

In general,

$$\overline{abc} = a \times 100 + b \times 10 + c$$
$$\overline{a_1 a_2 \cdots a_n} = a_1 \times 10^{n-1} + a_2 \times 10^{n-2} + \cdots + a_{n-1} \times 10 + a_n$$

Other Bases and Decimal Equivalents.

Other positive integers (other than 1) can be used as bases as well. Numbers in other bases are expressed in the same manner as the decimal (that is, base 10) system. For example, the binary system has base 2, and only uses two symbols (0 and 1) to represent numbers, and each of the binary digits also has their place values: units place, 2s place, 4s place, etc.

Usually, the base number is written as subscripts, such as 101_2. Here, for decimal numbers we omit the base number, that is, we simply write 345 instead of 345_{10}. Given a number expressed in other bases, we can find the equivalent values in decimal by adding up the place values.

- Binary (base 2) numbers: $1101_2 = 1 \times 2^3 + 1 \times 2^2 + 0 \times 2 + 1 = 8 + 4 + 1 = 13$
- Octal (base 8) numbers: $567_8 = 5 \times 8^2 + 6 \times 8 + 7 = 375$
- Hexadecimal (base 16) numbers: use A, B, C, D, E, F to represent 10, 11, 12, 13, 14, 15:

$$57C09AF_{16} = 5 \times 16^6 + 7 \times 16^5 + 12 \times 16^4 + 0 \times 16^3 + 9 \times 16^2 + 10 \times 16 + 15$$
$$= 92015023$$

Convert from decimal to binary:
- For the integer part: divide by 2, and record the remainder (either 0 or 1) at each step, and repeat. After reaching 0, write out the remainders in reverse order.

Example 2.1

Convert 13 to binary:
$$13 \div 2 = 6 \cdots 1$$
$$6 \div 2 = 3 \cdots 0$$
$$3 \div 2 = 1 \cdots 1$$
$$1 \div 2 = 0 \cdots 1$$

So $13_{10} = 1101_2$.

- For the fractional part: keep multiplying by 2 and taking the integer parts (also either 0 or 1) as the digits following the dot.

Example 2.2

Convert 0.6875 to binary:

$$
\begin{aligned}
0.6875 \times 2 &= 1.375 \quad \cdots \text{take} \quad 1 \\
0.375 \times 2 &= 0.75 \quad \cdots \text{take} \quad 0 \\
0.75 \times 2 &= 1.5 \quad \cdots \text{take} \quad 1 \\
0.5 \times 2 &= 1 \quad \cdots \text{take} \quad 1
\end{aligned}
$$

So $0.6875_{10} = 0.1011_2$.

Operations in other bases work the same way as in decimal system, just keep in mind that carry occurs at the base number (for example, in binary system, carry occurs whenever you have "2").

Example 2.3

In the decimal (base 10) system, $38 + 13 = 51$. Convert the summands to binary: $38 = 100110_2$, and $13 = 1101_2$. Let's perform binary addition (remember $1 + 1 = 10$ in the binary system):

$$
\begin{array}{ccccccc}
 & 1 & 0 & 0 & 1 & 1 & 0 \\
+ & & & 1 & 1 & 0 & 1 \\
\hline
 & 1 & 1 & 0 & 0 & 1 & 1
\end{array}
$$

The result 110011_2 is equivalent to 51 in base 10.

Problem 2.5 Convert:

(a) 11001111010110_2 to base 8

(b) 123_5 to base 4

(c) $ABCDEF_{16}$ to decimal

(d) 276 to base 16 (or "hexadecimal")

Problem 2.6 Find the value of base b such that the following addition is correct:

$$6651_b + 115_b = 10066_b$$

Problem 2.7 Find the pattern: In the sequence 110, 20, x, 11, y, z, 6, w, ... What are x, y, z, w?

Problem 2.8 Explain why: *Halloween = Christmas*

2.1.3 Divisibility

Notation: If an integer a divides an integer n evenly (that is, there is an integer c such that $ac = n$), then a is a factor (or divisor) of n, we write $a \mid n$ to denote that a divides n.

- A number $p > 1$ is prime if its only divisors are itself and 1.
- Every positive integer n has a unique factorization as a product of primes: $n = p_1^{e_1} p_2^{e_2} \cdots p_k^{e_k}$ for distinct primes p_i and $e_i > 0$.

Divisibility Rules: Let $n = \overline{a_k a_{k-1} \ldots a_1 a_0} = a_k 10^k + a_{k-1} 10^{k-1} + \cdots + a_1 10 + a_0, a_i \in \{0, 1, \ldots, 9\}$.

- By 2^j : If $\overline{a_{j-1} a_{j-2} \ldots a_1 a_0}$ is divisible by 2^j , then $2^j \mid n$.
- By 3: If $3 \mid (a_k + a_{k-1} + \cdots + a_1 + a_0)$, then $3 \mid n$.
- By 5: If $a_0 = 0$ or $a_0 = 5$, then $5 \mid n$.
- By 9: If $9 \mid (a_k + a_{k-1} + \cdots + a_1 + a_0)$, then $9 \mid n$.
- By 11: If $11 \mid (a_0 - a_1 + a_2 - \cdots + (-1)^k a_k)$, then $11 \mid n$.
- By 7 and 13: Use the fact that $1001 = 7 \times 11 \times 13$. Example: 15197. Split out the right-most 3 digits, so we have 15 and 197. Subtract the smaller from the larger, in this case $197 - 15 = 182$. Doing division, 182 is a multiple of both 7 and 13. Therefore 15197 is a multiple of both 7 and 13.

  For numbers with more than 6 digits: split the rightmost 3 digit in each step and do multiple steps.

Number of factors: For an integer n, $\sigma_0(n)$ is the number of factors of n. If $n = p_1^{e_1} p_2^{e_2} \cdots p_k^{e_k}$ is the prime factorization of n, then $\sigma_0(n) = \prod_{i=1}^{k} (e_i + 1)$.

Example 2.4

Consider the number $200 = 2^3 \times 5^2$. The exponents are 3 and 2, so $(3+1)(2+1) = 12$. Thus 200 has 12 factors. We can list them: $1, 2, 4, 5, 8, 10, 20, 25, 40, 50, 100, 200$. A better way is to list them in the following table:

	2^0	2^1	2^2	2^3
5^0	1	2	4	8
5^1	5	10	20	40
5^2	25	50	100	200

Greatest Common Divisors (GCD) and **Least Common Multiples (LCM)**: Let m and n be positive integers with prime factorizations $m = p_1^{e_1} p_2^{e_2} \cdots p_k^{e_k}$ and $n = p_1^{d_1} p_2^{d_2} \cdots p_k^{d_k}$ (here, e_i and d_i can equal zero).

(a) The greatest common divisor of m and n (denoted $\gcd(m,n)$) is the largest number d such that $d \mid m$ and $d \mid n$. $d = \prod_{i=1}^{k} p_i^{\min(d_i, e_i)}$.

(b) Two integers m and n are relatively prime if $\gcd(m,n)=1$.

(c) The least common multiple of m and n (denoted $\mathrm{lcm}(m,n)$) is the smallest number L such that $m \mid L$ and $n \mid L$. $L = \prod_{i=1}^{k} p_I^{\max(d_i,e_i)}$.

(d) Note $mn = \gcd(m,n) \times \mathrm{lcm}(m,n)$. (Prove it!)

(e) If $a \mid m$ and $a \mid n$, then $a \mid \gcd(m,n)$.

(f) If $m \mid k$ and $n \mid k$, then $\mathrm{lcm}(m,n) \mid k$.

Problem 2.9 In a store, The unit price of a certain item is an integer greater than \$1, and unchanged for two years. Last year the total sale of this item was \$36,963, and this year the total sale is \$59,570. What is the unit price?

Problem 2.10 Attach 3 digits after the number 503 so that the resulting 6-digit integer is a multiple of 7, 9 and 11.

Problem 2.11 Find a positive integer containing all ten digits: 0,1,2,3,4,5,6,7,8,9, that is a multiple of 126.

Problem 2.12 The integer $\overline{1a2a3a4a5a}$ is divisible by 11. What is a?

Problem 2.13 A seven-digit number has seven distinct digits, and it is divisible by 11. What is the largest such number?

Problem 2.14 A six-digit number, all of whose digits are distinct, is divisible by 11. Given that its left-most digit is 3. What is the smallest such number?

Problem 2.15 Let k be an even number. Is it possible to write 1 as the sum of the reciprocals of k odd integers?

Problem 2.16 Find all positive integers n for which $3n - 4$, $4n - 5$, and $5n - 3$ are all prime numbers.

Problem 2.17 (AIME 1984) The integer n is the smallest positive multiple of 15 such that every digit of n is either 0 or 8. Find n.

2.1.4 **Divisibility - continued**

> ### Theorem 2.1 Division Algorithm
>
> For any integers a and b, $a \neq 0$, there exists a unique pair (q, r) of integers such that $b = aq + r$ and $0 \leq r < |a|$. ♠

In the above theorem, the number q is called the *quotient*, and r is called the *remainder*.

> ### Theorem 2.2 Euclidean Algorithm
>
> Assume $m \geq n$ and let $r_0 = m$ and $r_1 = n$, then there exists sequences of integers $q_1, \ldots, q_k$ and $r_2, \ldots, r_k$ such that:
>
> $$
> \begin{aligned}
> r_0 &= r_1 q_1 + r_2 \quad \text{with} \quad 0 < r_2 < r_1 \\
> r_1 &= r_2 q_2 + r_3 \quad \text{with} \quad 0 < r_3 < r_2 \\
> &\;\;\vdots \qquad\qquad\qquad\qquad \vdots \\
> r_{k-1} &= r_k q_k \qquad\;\; \text{with} \quad r_k = \gcd(m, n)
> \end{aligned}
> $$
> ♠

The Euclidean Algorithm is used for finding the greatest common divisor of two integers. It is demonstrated in the following example.

> ### Example 2.5
>
> We find the greatest common divisor of 900 and 243 using the Euclidean algorithm:
>
> $$
> \begin{aligned}
> 900 &= 243 \times 3 + 171, \\
> 243 &= 171 \times 1 + 72, \\
> 171 &= 72 \times 2 + 27, \\
> 72 &= 27 \times 2 + 18, \\
> 27 &= 18 \times 1 + 9, \\
> 18 &= 9 \times 2. \quad \text{(Remainder=0, so the last divisor 9 is our answer)}
> \end{aligned}
> $$
>
> Therefore $\gcd(900, 243) = 9$. ♣

Theorem 2.3 Bézout's Identity

For any two positive integers m and n, there exist integers a and b such that $am + bn = \gcd(m,n)$.

Remark

This is an important theorem in number theory. The proof of this theorem is beyond the scope of this book. You may search online for further information about this theorem.

Problem 2.18 (IMO 1959) Show that the fraction

$$\frac{12n+1}{30n+2}$$

is irreducible for all positive integers n.

Problem 2.19 Determine the number of five-digit positive integers $\overline{abcde}$ (a,b,c,d, and e not necessarily distinct) such that the sum of the three-digit number $\overline{abc}$ and the two-digit number $\overline{de}$ (here d doesn't have to be nonzero) is divisible by 11.

Problem 2.20 The number 27000001 has exactly four prime factors. Find these primes factors.

Problem 2.21 (Putnam 2000) Given integers $n, m, n \geq m \geq 1$. Show that $\dfrac{\gcd(m,n)}{n} \dbinom{n}{m}$ is an integer.

Problem 2.22 (AHSME 1976) If p and q are primes and $x^2 - px + q = 0$ has distinct positive integral roots, find p and q.

Problem 2.23 Given a six-digit number $\overline{abcdef}$, whose digits are 1,2,3,4,5,6, not necessarily in this order. Assume that $6 \mid \overline{abcdef}$, $5 \mid \overline{abcde}$, $4 \mid \overline{abcd}$, $3 \mid \overline{abc}$, and $2 \mid \overline{ab}$. Find $\overline{abcdef}$.

Problem 2.24 Compute the product of all distinct positive divisors of 120^6 (express your answer as a power of 120).

2.2 Modular Arithmetic and Further Topics

2.2.1 Modular Arithmetic

Two numbers a and b are *congruent modulo m* (denoted $a \equiv b \pmod{m}$) if $m \mid (a-b)$; in other words, a and b have the same remainder when divided by m.

(a) If $a \equiv b \pmod{m}$ and $b \equiv c \pmod{m}$, then $a \equiv c \pmod{m}$.

(b) If $a \equiv b \pmod{m}$ and $c \equiv d \pmod{m}$, then $(a+c) \equiv (b+d) \pmod{m}$.

(c) If $a \equiv b \pmod{m}$ and $c \equiv d \pmod{m}$, then $ac \equiv bd \pmod{m}$.

Example 2.6

A typical example of modular arithmetic is on the calendar: day of the week. There are 7 days per week; use 0 for Sunday, 1 for Monday,..., 6 for Saturday. Suppose today is March 14, a Saturday. Then March 7, March 21, and March 28 are all Saturdays because the differences between those dates and today are multiples of 7. March 19 is 5 days from now, so it will be $6+5=11$, and $11 \equiv 4 \pmod{7}$, thus it is Thursday.

Example 2.7

The "Clock Arithmetic" is also an example of modular arithmetic. The modulus is 12. Suppose it is 11 o'clock now. 6 hours from now, $11+6 = 17 \equiv 5 \pmod{12}$, so it will be 5 o'clock (switched am/pm). If we want to calculate based on 24-hour time, then use modulo 24.

(d) If $a \equiv b \pmod{m}$ and $k \mid m$, then $a \equiv b \pmod{k}$.

(e) If (and only if) a and m are relatively prime, then there exists an integer $b < m$ such that $ba \equiv 1 \pmod{m}$. This is called the *modular multiplicative inverse* of a modulo m. This is similar to reciprocals of real numbers. In modular arithmetic, division is equivalent to multiplication by the inverse (if the inverse exists).

Proof. Consider the set of values $\{0, a, 2a, 3a, \ldots, (m-1)a\}$. When these values are divided by m, we show that the remainders are all distinct. In fact, suppose $ka \equiv la \pmod{m}$ for $0 \leq l \leq k < m$. By definition of congruence, $m \mid (k-l)a$. Since a and m are relatively prime, $m \mid (k-l)$. But we know that $0 \leq k-l < m$, so the only possibility is $k = l$. Thus all the m remainders mentioned above are distinct. But there are only m possible remainders modulo m. That means every

---

possible remainder is reached. In particular, one of the remainders must be 1. Let it be ba, then $ba \equiv 1 \pmod{m}$.   ∎

(f) If $a \equiv b \pmod{m}$, $d \mid a$, $d \mid b$, and $\gcd(d, m) = 1$, then $\dfrac{a}{d} \equiv \dfrac{b}{d} \pmod{m}$. (Prove it!)

Example 2.8

We know that $4 \equiv 18 \pmod{7}$. The numbers 4 and 18 have a common factor 2, and $\gcd(2, 7) = 1$, thus we can divide both numbers by 2 and get $2 \equiv 9 \pmod{7}$.

Remember that "Division is equivalent to multiplication by the inverse."

(g) More generally, if $a \equiv b \pmod{m}$, $d \mid a$, and $d \mid b$, then

$$\frac{a}{d} \equiv \frac{b}{d} \quad \left(\bmod \ \frac{m}{\gcd(d, m)}\right).$$

(Prove it!)

Example 2.9

Since $12 \equiv 40 \pmod{14}$, and 4 is a common factor of 12 and 40, we may cancel the 4, but $3 \not\equiv 10 \pmod{14}$. The reason is that $\gcd(4, 14) \neq 1$. But $\dfrac{14}{\gcd(4, 14)} = 7$, and we have $3 \equiv 10 \pmod{7}$.

Problem 2.25 Verify the following facts: Let n be an integer, then:
  (a) $n^2 \equiv 0$ or $1 \pmod{3}$;

  (b) $n^2 \equiv 0$ or $\pm 1 \pmod{5}$;

(c) $n^2 \equiv 0$ or 1 or 4 (mod 8);

(d) $n^3 \equiv 0$ or ± 1 (mod 9);

(e) $n^4 \equiv 0$ or 1 (mod 16);

Problem 2.26 For perfect squares, not all values in a certain moduli are possible remainders.

(a) Find the possible remainders of n^2 in (mod 4).

(b) Find the possible remainders of n^2 in (mod 9).

Problem 2.27 If $m > 1$ and $69 \equiv 90 \equiv 125 \pmod{m}$, what is m?

2.2.2 Modular Arithmetic - continued

Problem Solving Strategies:

- Parity (the property of being even or odd) analysis is often useful.
- If the squares or higher powers of integers are involved, it is often helpful to consider the remainders of the integer power in a particular modulus. (For example, in mod 3 and mod 4, squares have remainders 0 or 1; in mod 9, squares have remainders 0, 1, 4, or 7; in mod 16, 4th powers have remainders 0 or 1.)
- If the digits of a number are (or can be) rearranged, consider mod 9, because a number (in base 10) is congruent to the sum of its digits modulo 9. (Prove it!)

Residue Classes:

- A set S of integers is called a *complete set of residue classes modulo m* if for each $0 \le i \le m - 1$, there is an element $s \in S$ such that $i \equiv s \pmod{m}$. Also called a *complete residue system modulo m*.
  - For any integer a, $\{a, a+1, a+2, \ldots, a+m-1\}$ is a complete set of residue classes modulo m.
  - $\{0, 1, \ldots, m-1\}$ is the *minimal nonnegative complete set of residue classes modulo m*.
  - It is common to consider the complete set of residue classes

  $$\{0, \pm 1, \pm 2, \ldots, \pm k\}$$

  for $m = 2k+1$ and $\{0, \pm 1, \pm 2, \ldots, \pm(k-1), k\}$ for $m = 2k$.
- A set S of integers is called a *reduced set of residue classes modulo m* if for each $0 \le i \le m-1$ where $\gcd(i, m) = 1$, there is an element $s \in S$ such that $i \equiv s \pmod{m}$. Also called a *reduced residue system modulo m*.
  - If p is prime, $\{1, 2, \ldots, p-1\}$ is a reduced set of residue classes modulo p.
  - The number of elements in a reduced set of residue classes modulo m is denoted as $\phi(m)$. This is known as the *Euler ϕ function*, also called the *Euler totient function*. (This function will be explored in more details later.)

Example 2.10

Let $m = 10$, then $\{1, 3, 7, 9\}$ is a reduced set of residue classes modulo 10, and $\phi(10) = 4$.

We shall cover three important theorems. It is required to understand the proofs of the theorems, and know how the theorems are applied. It is not required to memorize the proofs.

Theorem 2.4 Fermat's Little Theorem

If p is prime and a is any integer, then $p \mid (a^p - a)$. Equivalently, if p does not divide a, then $a^{p-1} \equiv 1 \pmod{p}$. ♠

Proof. We apply a combinatorial method to prove that $p \mid (a^p - a)$. Let p be a prime and a be a positive integer such that $\gcd(a, p) = 1$ (the case where a is a multiple of p is trivial). Suppose we have beads of various colors and wish to make a bracelet consisting of p beads. Assume a is the number of the colors, and there are ample supply of beads of each color. If we do not rotate the bracelet, each position has a choices and the total number of possible bracelets is a^p. If we consider the bracelets whose only difference is a rotation as identical bracelets, then each bracelet is counted p times unless all of the beads are of the same color. The number of uni-color bracelets is a, thus $a^p - a$ must be divisible by p, which is what we wanted. ∎

A more conventional proof is given next.

Proof. Consider the reduced set of residue classes $a, 2a, 3a, \ldots, (p-1)a$ modulo p. This set covers all the non-zero remainders modulo p, thus

$$(a)(2a)(3a)\cdots((p-1)a) \equiv (p-1)! \pmod{p}.$$

So we get $(p-1)! \cdot a^{p-1} \equiv (p-1)! \pmod{p}$. Clearly $(p-1)!$ and p are relatively prime. Multiplying the modular multiplicative inverse of $(p-1)!$, we get $a^{p-1} \equiv 1 \pmod{p}$. ∎

Theorem 2.5 Wilson's Theorem

For every prime p, $(p-1)! \equiv -1 \pmod{p}$. ♠

Proof. The case $p = 2$ is trivial. For $p > 2$, we use a "pairing" method. For each a between 1 and $p-1$ inclusive, there is a modular multiplicative inverse b. Clearly the inverse of b is also a. Thus a and b is a pair such that $ab \equiv 1 \pmod{p}$. All numbers are paired up except for those values x where $x^2 \equiv 1 \pmod{p}$ (the only such x are 1 and $p-1$). So among $1, 2, 3, \ldots, p-1$, all are paired up except for 1 and $p-1$, so $(p-1)! \equiv p-1 \equiv -1 \pmod{p}$. ∎

Theorem 2.6 Chinese Remainder Theorem

Let $m_1,\ldots,m_k$ be pairwise relatively prime positive integers (that is, $\gcd(m_i,m_j)=1$ for all $i\neq j$). Let $b_1,\ldots,b_k$ be arbitrary integers. Then the system

$$\begin{aligned}
x &\equiv b_1 &&(\bmod\ m_1)\\
x &\equiv b_2 &&(\bmod\ m_2)\\
&\vdots &&\vdots\\
x &\equiv b_k &&(\bmod\ m_k)
\end{aligned}$$

has a unique solution modulo $m_1 m_2 \cdots m_k$.

For the Chinese Remainder Theorem, we shall not give a general proof, but use an example to illustrate how the solution is found, and the method is easily converted to a proof of the general case.

Example 2.11

(Based on ancient Chinese text.) An army general was counting his soldiers. The total number of soldiers was between 100 and 200. He let the soldiers stand in rows of 3 each. There were 2 soldiers left out. He changed the row size to 5 soldiers each, then there were 3 soldiers left out. Finally he changed to 7 soldiers per row, and there were 2 soldiers left out. How many soldiers were there?

Solution: This problem is equivalent to the following system of modular equations:

$$\begin{aligned}
x &\equiv 2 &&(\bmod\ 3)\\
x &\equiv 3 &&(\bmod\ 5)\\
x &\equiv 2 &&(\bmod\ 7)
\end{aligned}$$

In the Chinese text, the solution was described in a poem of 4 lines. The first line of the poem says (translated into modular arithmetic language): multiply the remainder mod 3 by **70**; in this case, $2\times 70=140$.
The second line says: multiply the remainder mod 5 by **21**; so $3\times 21=63$.
The third line says: multiply the remainder mod 7 by **15**; so $2\times 15=30$.
The fourth line says: add them all up and subtract a multiple of 105; that is $140+63+30-105=128$ (choose 128 because it was given that the number is between 100 and 200).
Thus there were 128 soldiers.

How did the solution work? Let's see what the three highlighted numbers (70, 21, and 15) are in the 3 moduli:

$$70 \equiv 1 \pmod{3} \; ; \; 21 \equiv 0 \pmod{3} \; ; \; 15 \equiv 0 \pmod{3}$$
$$70 \equiv 0 \pmod{5} \; ; \; 21 \equiv 1 \pmod{5} \; ; \; 15 \equiv 0 \pmod{5}$$
$$70 \equiv 0 \pmod{7} \; ; \; 21 \equiv 0 \pmod{7} \; ; \; 15 \equiv 1 \pmod{7}$$

If each number is multiplied by the corresponding remainder (as specified in the poem), we get

$$2 \times 70 \equiv 2 \pmod{3} \; ; \; 3 \times 21 \equiv 0 \pmod{3} \; ; \; 2 \times 15 \equiv 0 \pmod{3}$$
$$2 \times 70 \equiv 0 \pmod{5} \; ; \; 3 \times 21 \equiv 3 \pmod{5} \; ; \; 2 \times 15 \equiv 0 \pmod{5}$$
$$2 \times 70 \equiv 0 \pmod{7} \; ; \; 3 \times 21 \equiv 0 \pmod{7} \; ; \; 2 \times 15 \equiv 2 \pmod{7}$$

Therefore, when we add up the products,

$$2 \times 70 + 3 \times 21 + 2 \times 15 \equiv 2 \pmod{3}$$
$$2 \times 70 + 3 \times 21 + 2 \times 15 \equiv 3 \pmod{5}$$
$$2 \times 70 + 3 \times 21 + 2 \times 15 \equiv 2 \pmod{7}$$

And the remainders are unchanged if we add a multiple of $3 \times 5 \times 7 = 105$. Therefore the solution to the system of modular equations is

$$x \equiv 233 \equiv 23 \pmod{105}.$$

Remark

How do you apply this method to prove the general Chinese Remainder Theorem?

Problem 2.28 What is the remainder of $2^{50} + 3^{50}$ when divided by 13?

Problem 2.29 Show that there are no perfect squares in the sequence:

$$11, 111, 1111, 11111, \ldots$$

Problem 2.30 Is it possible to find two integers n and m such that $n^2 + m^2 = 2015$?

Problem 2.31 Can a 5-digit number consisting only of distinct even digits be a perfect square?

Problem 2.32 The number 2^{29} is a nine-digit number all of whose digits are distinct. Without computing the actual number, determine which of the ten digits is missing.

Problem 2.33 Show that in the set of 7! numbers consisting of the distinct permutations of the digits $1, 2, 3, 4, 5, 6, 7$, no member is a multiple of another.

Problem 2.34 A certain natural number n has a unit digit 9 when expressed in base 12. Find the remainder when n^2 is divided by 6.

Problem 2.35 Find the remainder when

$$37 + 377 + 3777 + 37777 + \cdots + 3777777777777777$$

is divided by 11 (note the last summand is 16 digits long (one 3 and fifteen 7's).

Problem 2.36 The Fibonacci sequence is defined by $F_1 = F_2 = 1$, and $F_{n+2} = F_{n+1} + F_n$, that is, the first two terms are both 1, and each subsequence term is the sum of the previous two terms. Fnd the remainder when F_{2011} is divided by 7.

Problem 2.37 How many zeros are there at the end of 1000! ?

Problem 2.38 There are two two-digit numbers whose square ends in the same two-digit number. Find them.

2.2.3 Pythagorean Triples

A *Pythagorean triple* is a triple of positive integers (a,b,c) such that $a^2+b^2=c^2$.

> **Example 2.12**
>
> Some well-known Pythagorean triples are: $(3,4,5)$, $(5,12,13)$, $(7,24,25)$, $(8,15,17)$. ♣

If a triple of integers (a,b,c) is a Pythagorean triple, then so is (ka,kb,kc) where k is any positive integer.

> **Example 2.13**
>
> Since $(3,4,5)$ is a Pythagorean triple, $(6,8,10)$, $(9,12,15)$, $(12,16,20)$ are all Pythagorean triples. ♣

A *primitive Pythagorean triple* is a Pythagorean triple where $\gcd(a,b,c)=1$. All Pythagorean triples in Example 2.2.3 are primitive ones.

All primitive Pythagorean triples (a,b,c) can be obtained from the formula: $a=m^2-n^2, b=2mn, c=m^2+n^2$ where m,n are integers such that $\gcd(m,n)=1$.

Problem 2.39 Find all Pythagorean triples containing the number 29.

Problem 2.40 Find all Pythagorean triples containing the number 15. (Note: there are 5 such triples.)

Problem 2.41 Are there any prime numbers between 2020 and 2030?

Problem 2.42 An 8-digit number $\overline{141A28B3}$ is a multiple of 99. Find A and B.

Problem 2.43 Find the smallest six digit integer with leading digit 7, and all digits are distinct, that is divisible by 11.

Problem 2.44 Find the smallest positive integer satisfying both of the following requirements:

   (a) Its units digit is 6;
   (b) If the units digit 6 is moved before the first digit, the new number is 4 times the original number.

Problem 2.45 Find the smallest positive multiple of 225, all of whose digits are 0 or 1 in its base 10 representation.

2.2 Modular Arithmetic and Further Topics

Problem 2.46 Find the smallest positive integer n such that $\sqrt{2000n}$ is an integer.

Problem 2.47 Given that $2^{96} - 1$ is divisible by two integers between 60 and 70. What are these two integer?

Problem 2.48 Find the greatest common divisor of the following ten integers: $2000^3 + 3 \cdot 2000^2 + 2 \cdot 2000$, $2001^3 + 3 \cdot 2001^2 + 2 \cdot 2001$, $\ldots$, $2008^3 + 3 \cdot 2008^2 + 2 \cdot 2008$, $2009^3 + 3 \cdot 2009^2 + 2 \cdot 2009$.

Problem 2.49 The six-digit number $\overline{xy342z}$ is divisible by 396. Find all such numbers.

Problem 2.50 Find the largest multiple of 11 among the nine-digit numbers, whose digits are all distinct.

Problem 2.51 Find all numbers n less than 50 with the following property: the product of the divisors of n is equal to n^2.

Problem 2.52 (1984 ARML I-7) Find all positive integers n less than twenty such that $49 \mid n! + (n+1)! + (n+2)!$.

Problem 2.53 What is the largest natural number k such that $\dfrac{1001 \cdot 1002 \cdots 2000}{11^k}$ is an integer?

Problem 2.54 What is the remainder when 9^{2012} is divided by 11?

Problem 2.55 What are the last two digits of 2012^{2012}?

Problem 2.56 A number m is the smallest positive integer that gives remainder 1 when divided by 3, remainder 5 when divided by 7, and remainder 4 when divided by 11. What is the remainder when m is divided by 4?

Problem 2.57 The Fibonacci sequence is defined by $F_1 = F_2 = 1$, and $F_{n+2} = F_{n+1} + F_n$, that is, the first two terms are both 1, and each subsequent term is the sum of the previous two terms. Find the remainder when F_{2010} is divided by 7.

Problem 2.58 A four digit number minus the sum of its digits, the result is $\overline{20d0}$. What is d?

Problem 2.59 A five digit number $\overline{4a77b}$ is divisible by 99, find the values of a, b.

Problem 2.60 A two digit number equals the sum of its tens digit and the square of its units digit. What is this two digit number?

Problem 2.61 (1980 Canada) Let $\overline{a679b}$ be a five-digit number. If $72 \mid \overline{a679b}$, find the values of a and b.

Problem 2.62 Let n be a positive integer, such that $n+3$ is a multiple of 5, and $n-3$ is a multiple of 6. Find the smallest such n.

Problem 2.63 Factorial with trailing zeros.

   (a) (1985 NYSML T-3) For how many positive integral values of n does $n!$ end with precisely 25 zeros? What are they?

   (b) Same question, but what if $n!$ is represented in base eight?

2.2.4   More Challenge Problems

Further practices on what we have learned so far. Some of these are challenging; please pay special attention to the techniques used when solving these problems. The techniques include (but not limited to these):

- Use modular arithmetic to reduce the numbers whenever possible in calculations related to remainders.
- Find patterns of remainders.
- Use the properties of special moduli:
    - In mod 3 or mod 4, squares only has value 0 or 1.
    - In mod 9, the sum of the digits has the same remainder as the original number (Use mod 9 if the order of digits is unimportant).
- Try some small numbers first before doing the general case.
- Use algebraic formulas to factor the given expressions.

Make sure to write down the process of your solutions.

Problem 2.64 Find the remainder when 1996^{2000} is divided by 29.

Problem 2.65 Find the remainder when 2001^{2000} is divided by 49.

Problem 2.66 Find the remainder when $2222^{5555} + 5555^{2222}$ is divided by 7.

Problem 2.67 Let x, y be positive integers, $x < y$, and $x + y = 667$. Given that $\dfrac{\text{lcm}(x,y)}{\gcd(x,y)} = 120$. Find all such pairs (x, y).

Problem 2.68 Find all ordered pairs of positive integers (x, y) such that $1! + 2! + 3! + \cdots + x! = y^2$.

Problem 2.69 In the Cartesian coordinate system, how many grid points (x, y) satisfy $(|x| - 1)^2 + (|y| - 1)^2 < 2$?

Problem 2.70 Find all possible positive integers n such that $323 \mid 20^n + 16^n - 3^n - 1$.

Problem 2.71 Let p and d be positive integers, and $6 \nmid d$. Assume that $p, p+d, p+2d$ are all prime numbers. Then $p + 3d$ must be ... (select all that apply)

(A) prime   (B) multiple of 9   (C) multiple of 3   (D) either prime or multiple of 9.

Problem 2.72 Find all possible positive integers n such that $2^n - 1$ is a multiple of 7.

Problem 2.73 Let A be the sum of the digits of 5^{10000}, and B be the sum of digits of A, and C be the sum of digits of B. What is C?

Problem 2.74 Given a sequence: 1,4,8,10,16,19,21,25,30,43. If a group of consecutive terms in this sequence has a sum that is a multiple of 11, then call this group a "fine" group. How many "fine" groups are there?

Problem 2.75 In $\triangle ABC$, all three sides have integer lengths. Assume that $AB = 21$, and its perimeter is 54. Also known that its area is a positive integer. What are BC and CA?

Problem 2.76 Given a positive integer n, how many positive integers a are there such that $n^6 + 3a$ is a perfect cube? Justify your answer.

Problem 2.77 From the set $\{1, 2, \ldots, 100\}$, select k numbers. What is the minimum value of k such that it is guaranteed to have two numbers that are not relatively prime?

Problem 2.78 For positive integer k, let $M = 2(2k - 1)$, which of the following must be true?

   (a) M is not a perfect square for any k.
   (b) There are infinitely many k such that M is a perfect square.
   (c) There is a unique k such that M is a perfect square.
   (d) There are finitely many, but more than 1, values of k such that M is a perfect square.

Problem 2.79 Put a positive or negative sign in front of each of $1, 2, 3, \ldots, 2003$, and take the sum, then this sum must be
(A) Odd   (B) Even   (C) multiple of 3   (D) none of those

Problem 2.80 Given that a, b, n are positive integers. Assume that for any positive integer $k \neq b$, $(k-b) \mid (k^n - a)$, then which of the following must be true?
(A) $a > b^n$ (B) $a < b^n$ (C) $a = b^n$ (D) It depends.

Problem 2.81 Find all ordered triples (x, y, z) of prime numbers satisfying equation $x(x+y) = z + 120$.

Problem 2.82 Three distinct positive integers a, b, c are pairwise relatively prime, and the sum of any two is a multiple of the third one. What is the product abc?

Problem 2.83 Given that a and b are both prime numbers and $p = a^b + b^a$ is also prime. What is p?

Problem 2.84 Let n be a positive integer, and $\dfrac{n(n+1)}{2} - 1$ is a prime number. Find all possible values of n.

2.3 Number Theory Techniques

In this section we shall learn several categories of methods for solving number theory problems.

2.3.1 Problem Solving Techniques in Number Theory

- **Expressing Integers in Different Ways**

  Common forms of expressions are:
  - Place values: $n = a_k \cdot 10^k + a_{k-1} \cdot 10^{k-1} + \cdots + a_1 \cdot 10 + a_0$;
  - Division with a remainder: $n = mq + r$, where q is the quotient and r is the remainder, and $r < |m|$;
  - Prime factorization: $n = p_1^{e_1} p_2^{e_2} \cdots p_k^{e_k}$;
  - Power of 2 times an odd number: $n = 2^m \cdot t$ where t is an odd number.

- **Enumeration**

  The enumeration method can be described as "smart brute force." Case analysis is one example. Common methods of Enumeration include categorizing based on residue classes, parity, or values.

- **Finding Patterns**

  One very common and effective method is to start with solving a few similar problems with smaller sizes, and find a pattern.

- **Construction**

  In number theory, we can construct some special structures, number or set of numbers with special properties to solve a problem.

- **Contradiction**

  Make an assumption opposite the desired conclusion, then use correct logical reasoning to reach a contradiction, and that proves the original conclusion is correct.

- **Pairing**

  Legend has it that Gauss invented the pairing method at age 8 when he was adding up $1 + 2 + \cdots + 100$. Pairing has a lot of different applications.

- **Estimation**

  Use inequalities to enlarge or reduce a quantity to find its range of values and essential characteristics, and narrow down the choices for analysis. In number theory, there are at most finitely many integers in a given finite range; therefore it is beneficial to estimate and narrow down the choices and check each case.

- **Parity Analysis**

  All integers can be divided into two categories: even and odd. Some properties of even and odd numbers are quite obvious, and yet very useful in solving certain problems. For example, an even number is never equal to an odd number. Or, the sum of an odd number of odd numbers is an odd number.

2.3.2 **Practice Problems**

Problem 2.85 Given four cards with red, yellow, white, and blue colors, each card having a digit on it. Mike put the cards in a row in the order of red, yellow, white, and blue, to form a four digit number. Then he calculated the difference between this four digit number and 10 times the sum of its digits. He found out that no matter what digit was on the white card, the result of the calculation was always 1998. What are the digits on the red, yellow, and blue cards?

Problem 2.86 In a mathemagic show, the mathemagician asked Nick (a person he picked from the audience) to (1) think about a three digit number $\overline{abc}$; and (2) write down five numbers: $\overline{acb}$, $\overline{bac}$, $\overline{bca}$, $\overline{cab}$, $\overline{cba}$; and (3) add up these five numbers to get N. As soon as Nick said the value of N, the mathemagician announced the original number $\overline{abc}$. If $N = 3194$, what was $\overline{abc}$?

Problem 2.87 From natural numbers $1, 2, 3, \ldots, 1000$, at most how many can be selected such that the sum of any three of the selected numbers is a multiple of 18?

Problem 2.88 Find positive integer n that is divisible by both 5 and 49 and has exactly 10 positive divisors.

Problem 2.89 Let N be the least common multiple of $1, 2, 3, \ldots, 1998, 1999, 2000$, and 2^k be the maximum power of 2 that divides N. What is k?

Problem 2.90 Partition the first n positive integers into several non-intersecting subsets, so that none of the subsets contain both m and $2m$ for any m. At least how many subsets should there be?

Problem 2.91 Find all three digit number n such that the remainder when n is divided by 11 is equal to the sum of the squares of n's digits.

Problem 2.92 Attach a positive integer N to the right of any positive integer (for example, attaching 8 to the right of 57, we get 578), if the new number is always divisible by N no matter what the other positive integer is, then call N a "magic number". Find all "magic numbers" less than 2000.

Problem 2.93 Given three cards, each with an integer between 1 and 10 (inclusive) on it. After shuffling, deal the three cards to Adam, Bob, and Chris, one card each. Everyone write down the number on his card, and repeat the process: shuffle, deal, record. After some rounds, each person adds up the numbers he received. The sums are 13, 15, and 23. What are the numbers on the three cards?

Problem 2.94 Is it possible to express 99^{99} as the sum of 99 consecutive odd positive integers? How about 99! ?

Problem 2.95 From the numbers $1, 2, 3, \ldots, 999$, cross out the least possible number of numbers so that none of the remaining numbers is the product of two other remaining numbers. Which numbers should be crossed out?

Problem 2.96 Is there a 3 digit number $\overline{abc}$ such that $\overline{abc} = \overline{ab} + \overline{bc} + \overline{ac}$?

Problem 2.97 Given any 17-digit number, reverse its digits to get another number, and then add the new number and the original number. Show that at least one digit in the sum is even.

Problem 2.98 A magic coin machine behaves as follows. If you put in a penny, it returns a dime and a nickel. If you put in a nickel, it returns 4 dimes. If you put in a dime, it returns 3 pennies. Becky started with a penny and a nickel, and kept putting coins into the machine and collected the returned coins. Is it possible at some point of time that the number of pennies Becky has is exactly 10 less than the number of dimes?

Problem 2.99 The 3×3 table below contains 9 primes numbers. Define an "operation" as adding the same positive integer to the 3 numbers in one row or one column. Is it possible to change all numbers in the table to the same number after several operations?

2	3	5
13	11	7
17	19	23

Problem 2.100 Find the sum of all the digits in the numbers $1, 2, 3, \ldots, 9999999$.

Problem 2.101 A department store distributes 9999 raffle tickets to the customers, each ticket has a 4-digit number from 0001 to 9999. If the sum of the first two digit equals the sum of the last two digits, then the ticket is called a "lucky ticket". For example, ticket number 0945 is a lucky ticket. Show that the sum of all the numbers on the lucky tickets is divisible by 101.

Problem 2.102 Let

$$\frac{m}{n} = 1 + \frac{1}{2} + \frac{1}{3} + \cdots + \frac{1}{88}$$

where $\gcd(m,n) = 1$. Show that $89 \mid m$.

Problem 2.103 One integer n equals the sum of 4 distinct fractions of form $\dfrac{m}{m+1}$ (m is a positive integer). Find this integer n, and also find at least one such set of 4 fractions that add up to n.

Problem 2.104 Find the largest n such that $n!$ has exactly 106 zeros at the end.

Problem 2.105 A book has 192 pages and is printed double-sided on 96 sheets of paper. Each page has its page number printed at a corner. Kiran tore 25 sheets out of the book, and added up all the page numbers printed on these sheets. Is it possible that the sum is 2010?

Problem 2.106 The Gauss Middle School has 98 students, each has a unique student number, from 1 through 98. Is it possible to let the students stand in several rows, such that there is a student in each row whose number equals the sum of the numbers of the rest of the students in the same row?

Problem 2.107 Ninety-nine students participated in the Planetary Math Olympiad. There are 30 questions in the PMO, and the scoring is as follows. There are 15 base points; add 5 for each correct answer, subtract 1 for each incorrect answer, and add 1 for each unanswered question. If all the scores were added up, is the sum an even number or an odd number?

Problem 2.108 Seventy-seven coins are put on the table, showing heads. First turn over all 77 coins. The second step, turn over 76 of them. The third step, turn over 75 of them, and so on. The 77th step, only turn over 1 of the coins. Is it possible to make all 77 coins show tails? If not, explain why. If yes, describe how it can be done.

Problem 2.109 The difference between two positive integers is multiplied by their product; can the final product be 45045?

Problem 2.110 A certain 4-digit number satisfy the following: its tens digit minus 1 equals its units digit; the units digit plus 2 equals the hundreds digit; and if the digits of this 4-digit number is reversed, the new number plus the original number equals 9878. Find the original 4-digit number.

Problem 2.111 Let a,b,c,d be a permutation of the numbers $1,2,3,4$, satisfying $a < b, b > c, c < d$, and $\overline{abcd}$ is a 4-digit number. Find all such 4-digit numbers.

Problem 2.112 Let n be the smallest multiple of 75 that has exactly 75 factors. Find $\dfrac{n}{75}$.

Problem 2.113 Maggy has a deck of 100 cards. She starts with the card on top, and do the following: throw away the top card, and put the next top card to the bottom; then throw away the new top card, and put the next top card at the bottom, and so on, until only one card is left. Which card from the original deck is the remaining card?

Problem 2.114 What is the largest even number that cannot be written as the sum of two odd composite numbers?

Problem 2.115 (AIME 1983) Find the largest two-digit prime factor for the integer $\binom{200}{100}$.

Problem 2.116 (Kiev 1973) Find three primes numbers whose product is five times their sum.

Problem 2.117 (Leningrad 1980) Let p and q be primes. The equation $x^4 - px^3 + q = 0$ has an integer root. Find the values of p and q.

Problem 2.118 (Kiev 1978) Find the smallest positive integers a and $b(b > 1)$, such that $\sqrt{a\sqrt{a\sqrt{a}}} = b$.

Problem 2.119 Arrange the numbers $1, 2, 3, \ldots, 999$ on a circle, in that order. Start from 1, do the following: skip 1, cross out 2 and 3; skip 4, cross out 5 and 6. Each step skip one number and cross out the next two. Which number is the last one remaining?

Problem 2.120 Form 4-digit numbers with the digits $0, 1, 2, 3, 4$, with no repeating digits within each number (for example, 1023, 3412). Find the sum of all such 4-digit numbers.

Problem 2.121 Twenty-seven countries send delegations to an international conference, each country has two representatives. Is it possible to arrange the 54 people around a round table, so that between the two people from any country, there are 9 people from other countries?

Problem 2.122 A magic square is a square matrix with the property that the sums of the numbers on each row, column, and diagonal are the same. This sum is called the "magic sum". Is it possible that a 3×3 magic square has a magic sum 1999?

Problem 2.123 Express the fractions $\dfrac{7}{332}$ and $\dfrac{1949}{1999}$ in the form $\dfrac{1}{m} + \dfrac{1}{n}$, where m, n are positive integers. If not possible, explain why.

Problem 2.124 A five-digit number N consists of 5 distinct nonzero digits, and N equals the sum of all possible 3-digit numbers made up of 3 of its 5 digits. Find all such 5-digit numbers N.

Problem 2.125 Evaluate:

$$\left\lfloor \frac{199 \times 1}{97} \right\rfloor + \left\lfloor \frac{199 \times 2}{97} \right\rfloor + \cdots + \left\lfloor \frac{199 \times 96}{97} \right\rfloor.$$

Problem 2.126 Given plenty of apples, pears, and oranges, all mixed together in one pile. You want to separate the fruits into several piles, each containing all three kinds of fruits, to guarantee that you can select two piles that, when combined together, there are even number of each kind of fruits in the combined pile. How many piles should you separate the fruits into?

Problem 2.127 Can the number 1010 be expressed as the sum of 10 consecutive integers?

Problem 2.128 How many even numbers are among the first 100 Fibonacci numbers?

Problem 2.129 Assume that in a chess game, the winner earns 1 point, the loser gets −1, and both get 0 if it is a draw. In a tournament among several students, each student played one game against every other student. Given that one of the students received a total of 7 points, and another student received 20 points. Show that there was at least one draw during the games.

Problem 2.130 There are the 909 numbers on the board, $1, 2, \ldots, 909$. Each step, erase any two numbers from the board and write their nonnegative difference onto the board, until there is only one number left. Is this last number even or odd?

Problem 2.131 (Putnam 1989) Let K be the set of all positive integers consisting of alternating digits 1 and 0: $\{1, 101, 10101, 1010101, \ldots\}$. Which elements of K are prime numbers?

3. Algebra

3.1 Factoring Polynomials

Polynomial factorization is one of the most fundamental skills in algebra. Factorization is the inverse operation of polynomial multiplication. We shall introduce the basic methods for polynomial factorization. In this exercise, we only consider the case where all coefficients are real numbers.

For each problem, it is required to factor the polynomial completely, so that any portion of the resulting expression cannot be factored any further.

3.1.1 Find Common Factors

The first step in any factoring problem is to search for common factors. Factoring out the common factor is the backward application of the Distributive Law: $ab + ac = a(b + c)$. Here a, b, c can be numbers or expressions. Sometimes it takes some work in order to get the common factor to appear.

Example 3.1

Factor the following.
(a) $10x^{10}y^8 + 5x^5y^9$.
   Solution: $5x^5y^8(2x^5 + y)$
(b) $(2x + 3y)^3 - 8x^3 - 27y^3$.
   Solution: $18xy(2x + 3y)$

Problem 3.1 Factor $10x^2y^2 - 15xy^3 + 25xy^2z$.

Problem 3.2 Factor $6x(a-b)^4 - 30x(b-a)^3$.

3.1.2 Use formulas

Commonly used formulas:

$$
\begin{aligned}
a^2 - b^2 &= (a+b)(a-b) \\
a^2 \pm 2ab + b^2 &= (a \pm b)^2 \\
a^3 \pm 3a^2b + 3ab^2 \pm b^3 &= (a \pm b)^3 \\
a^3 + b^3 &= (a+b)(a^2 - ab + b^2) \\
a^3 - b^3 &= (a-b)(a^2 + ab + b^2) \\
a^2 + b^2 + c^2 + 2ab + 2bc + 2ca &= (a+b+c)^2 \\
a^3 + b^3 + c^3 - 3abc &= (a+b+c)(a^2 + b^2 + c^2 - ab - bc - ca)
\end{aligned}
$$

$$
\begin{aligned}
a^n - b^n &= (a-b)(a^{n-1} + a^{n-2}b + a^{n-3}b^2 + \cdots + ab^{n-2} + b^{n-1}), & n \in \mathbb{N} \\
a^n - b^n &= (a+b)(a^{n-1} - a^{n-2}b + a^{n-3}b^2 - \cdots + ab^{n-2} - b^{n-1}), & n \in \mathbb{N} \text{ even} \\
a^n + b^n &= (a+b)(a^{n-1} - a^{n-2}b + a^{n-3}b^2 - \cdots - ab^{n-2} + b^{n-1}), & n \in \mathbb{N} \text{ odd}
\end{aligned}
$$

Example 3.2

Prove the formula for factoring $a^3 + b^3 + c^3 - 3abc$.
Solution:

$$
\begin{aligned}
&a^3 + b^3 + c^3 - 3abc \\
={}& (a+b)^3 - 3ab(a+b) + c^3 - 3abc \\
={}& (a+b)^3 + c^3 - 3ab(a+b+c) \\
={}& [(a+b+c)(a+b)^2 - c(a+b) + c^2] - 3ab(a+b+c) \\
={}& (a+b+c)(a^2 + b^2 + c^2 - ab - bc - ca)
\end{aligned}
$$

Note: A second method to prove this formula would be to expand the factored form and see if the result is identical to the original polynomial.

Example 3.3

Factor $y^6 - y^3$.
Solution:

$$
\begin{aligned}
y^6 - y^3 &= y^3(y^3 - 1) \\
&= y^3(y-1)(y^2 + y + 1)
\end{aligned}
$$

Note: We could also do $y^6 - y^3 = (y^2)^3 - y^3$ and apply the difference of cubes formula. However, finding common factors first is usually simpler.

Problem 3.3 Factor the following:

(a) $-2x^{5n-1}y^n + 4x^{3n-1}y^{n+2} - 2x^{n-1}y^{n+4}$

(b) $x^3 - 8y^3 - z^3 - 6xyz$

(c) $a^2 + b^2 + c^2 - 2bc + 2ca - 2ab$

(d) $a^7 - a^5b^2 + a^2b^5 - b^7$

Problem 3.4 Factor $a^{32} - b^{32}$.

Problem 3.5 Factor $x^{15} + x^{14} + x^{13} + \cdots + x^2 + x + 1$.

3.1.3 The UNFOIL Method

The UNFOIL method is the reverse of the FOIL method: $abx^2 + (ad+bc)x + cd = (ax+c)(bx+d)$.

Example 3.4

Factor $22y^2 - 35y + 3$.
Solution: Use some trial and error to get $22 = 2 \times 11$ and $3 = (-3)(-1)$, and $-35 = 2(-1) + 11(-3)$, thus

$$22y^2 - 35y + 3 = (2y-3)(11y-1).$$

This method can be shown in a cross-multiplication manner.

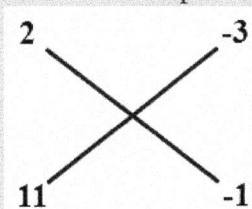

In the diagram, the leading coefficient 22 is factored as 2×11 and written in the first column; the constant 3 is factored as $(-3)(-1)$ and written in the second column. The cross-multiplication and sum $2(-1) + 11(-3) = -35$ is exactly equal to the middle term's coefficient, so this method is successful (it usually takes a few trial and error steps to find the correct arrangement of the numbers), and each row contains the coefficients of one factor: $2y-3$ and $11y-1$.

Problem 3.6 Factor the following:

(a) $2x^2 - 7x + 3$.

(b) $3k^2 - 5k - 2$.

(c) $2p^2 + p - 3$.

3.1.4 Factor by Grouping

If there are 4 or more terms, we can group some terms together and factor the portions first, and sometimes the portions end up with common factors or fit into a formula, then factoring is possible.

Example 3.5

Factor $xy + xz + yw + wz$.
Solution:
$$
\begin{aligned}
xy + xz + yw + wz &= (xy + xz) + (yw + wz) \\
&= x(y + z) + w(y + z) \\
&= (x + w)(y + z).
\end{aligned}
$$

Example 3.6

Factor $a^4 + a^3 + a^2b + ab^2 + b^3 - b^4$.
Solution:
$$
\begin{aligned}
& a^4 + a^3 + a^2b + ab^2 + b^3 - b^4 \\
={}& (a^4 - b^4) + (a^3 + b^3) + (a^2b + ab^2) \\
={}& (a+b)(a-b)(a^2+b^2) + (a+b)(a^2-ab+b^2) + (a+b)ab \\
={}& (a+b)[(a^2+b^2)(a-b) + (a^2-ab+b^2) + ab] \\
={}& (a+b)[(a^2+b^2)(a-b) + (a^2+b^2)] \\
={}& (a+b)(a^2+b^2)(a-b+1).
\end{aligned}
$$

3.1.5 Split and add terms

Split one term to two or more terms; add two terms that cancel each other. These techniques can enable the polynomial to be factored by grouping.

Example 3.7

Factor $x^3 - 9x + 8$.
Solution: We can get several solutions by splitting and adding different terms.
First Solution: $x^3 - 9x + 8 = x^3 - 9x - 1 + 9 = (x^3 - 1) - 9(x - 1) = (x - 1)(x^2 + x + 1) - 9(x - 1) = (x - 1)(x^2 + x + 1 - 9) = (x - 1)(x^2 + x - 8)$.
Second Solution: $x^3 - 9x + 8 = x^3 - x - 8x + 8 = (x^3 - x) - (8x - 8) = x(x + 1)(x - 1) - 8(x - 1) = (x^2 + x - 8)(x - 1)$.
Third Solution: $x^3 - 9x + 8 = 9x^3 - 8x^3 - 9x + 8 = (9x^3 - 9x) - (8x^3 - 8) = 9x(x + 1)(x - 1) - 8(x - 1)(x^2 + x + 1) = (x - 1)(9x^2 + 9x - 8x^2 - 8x - 8) = (x - 1)(x^2 + x - 8)$.
Fourth Solution: $x^3 - 9x + 8 = x^3 - x^2 + x^2 - 9x + 8 = (x^3 - x^2) + (x^2 - 9x + 8) = x^2(x - 1) + (x - 8)(x - 1) = (x^2 + x - 8)(x - 1)$.

Example 3.8

Sophie Germain's technique. Factor the following: $x^4 + 4y^4$.
Solution: $x^4 + 4y^4 = x^4 + 4x^2y^2 + 4y^4 - 4x^2y^2 = (x^2 + 2y^2)^2 - (2xy)^2 = (x^2 + 2xy + 2y^2)(x^2 - 2x^2y^2 + 2y^2)$.

Problem 3.7 Factor the following:
 (a) $x^9 + x^6 + x^3 - 3$.

 (b) $(m^2 - 1)(n^2 - 1) + 4mn$.

(c) $(x+1)^4 + (x^2-1)^2 + (x-1)^4$.

(d) $a^3b - ab^3 + a^2 + b^2 + 1$.

3.2 Solving Equations

Problems involving algebraic equations are very common in math contests. These problems usually require highly non-routine techniques. Therefore, in addition to grasping the basic concepts and basic methods, we should always develop new thinking skills and learn new tricks of solving equations.

3.2.1 Quadratic Equations: Discriminant and Vieta's Theorem

A quadratic equation is an equation in x of the form $ax^2 + bx + c = 0$ where $a \neq 0$. If the coefficients a, b, c are given, the roots (solutions) are determined as well. On the other hand, if the roots are given, the coefficients a, b, c are also determined up to a common factor (that is to say, if a, b, c are multiplied by the same nonzero number, the roots don't change). This shows that in a quadratic equation, the roots and the coefficients have very close relationship. This relationship is expressed in the discriminant and Vieta's Theorem.

Roots of A Quadratic Equation

There are several ways to solve a quadratic equation. In general, the roots x_1 and x_2 are given by the quadratic formula:

$$x_{1,2} = \frac{-b \pm \sqrt{b^2 - 4ac}}{2a}.$$

You should know how to derive the quadratic formula.

The Discriminant

Note that the expression $b^2 - 4ac$ is inside the square root. Therefore the solution for the quadratic equation depends on the sign of this expression. Denote $\Delta = b^2 - 4ac$. This is called the *discriminant*. There are three possibilities:

1) If $\Delta > 0$, the equation has two distinct real roots;
2) If $\Delta = 0$, the equation has exactly one real root; or we say it has two identical roots, or double roots.
3) If $\Delta < 0$, the equation has no real roots.

Example 3.9

Determine the number of roots without solving the equation:
   (a) $3x + 4x - 5 = 0$.
      Solution: $\Delta = 4^2 - 4(3)(-5) = 16 + 60 = 76 > 0$, thus there are two real roots.
   (b) $4x^2 + 20x + 25 = 0$.
      Solution: $\Delta = 20^2 + 4(4)(25) = 400 - 400 = 0$, so there is one real root.
   (c) $2x^2 + 2x + 3 = 0$.
      Solution: $\Delta = 2^2 - 4(2)(3) = 4 - 24 = -20 < 0$, so there are no real roots.

Example 3.10

(2005 AMC 10A/12A) There are two values of a for which the equation $4x^2 + ax + 8x + 9 = 0$ has only one solution for x. What is the sum of those values of a?

Solution: For the quadratic equation to have only one solution, the discriminant has to be 0. Thus $\Delta = (a+8)^2 - 4 \cdot 4 \cdot 9 = 0$. Therefore $(a+8)^2 = 144$, and then $a + 8 = \pm 12$, which means $a = 4$ or $a = -20$. So the answer is $4 + (-20) = -16$.

Vieta's Theorem

   Vieta's Theorem (or Vieta's Formulas) shows the relation between the roots and the coefficients.

Theorem 3.1 Vieta's Formulas

If x_1 and x_2 are the roots for the quadratic equation $ax^2 + bx + c = 0 (a \neq 0)$, then
$$x_1 + x_2 = -\frac{b}{a}, \quad \text{and} \quad x_1 x_2 = \frac{c}{a}.$$

These relations can be easily proved using the quadratic formula.
Note: The converse of this theorem is also true.

Using the discriminant and Vieta's Theorem, we can determine some properties of the roots based on the coefficients, or find out the range of parameters appearing in the coefficients based on the properties of the roots, etc. Using the discriminant and Vieta's Theorem, we can solve many algebra, geometry, and trigonometry, and some more complicated problems. In the AMC 10 and 12, many problems can be easily solved using discriminant or Vieta's Theorem.

Example 3.11

(2005 AMC 10B/12B) The quadratic equation $x^2 + mx + n = 0$ has roots that are twice those of $x^2 + px + m = 0$, and none of $m, n,$ and p is zero. What is the value of n/p?

Solution: Assume the two roots of the latter equation are x_1 and x_2, then the roots of the former equation are $2x_1$ and $2x_2$. By Vieta's Theorem, $x_1 + x_2 = -p$, $x_1 x_2 = m$, and also $2x_1 + 2x_2 = -m$ and $(2x_1)(2x_2) = n$. So we get $m = 2p, 4m = n$. So $n/p = 8$.

Example 3.12

(2008 AMC 10B) A quadratic equation $ax^2 - 2ax + b = 0$ has two real solutions. What is the average of the solutions?

Solution: 1. By Vieta's Theorem, $x_1 + x_2 = \dfrac{2a}{a} = 2$, So the average $\dfrac{x_1 + x_2}{2} = 1$.

Example 3.13

Find all ordered pairs (a, b) such that $a^2 + b^2$ is prime, and the equation $x^2 + ax + 1 = b$ has two positive integer roots.

Solution: No such pairs exist. Let x_1 and x_2 be the two positive integer roots, then $x_1 + x_2 = -a$ and $x_1 x_2 = 1 - b$, thus $a^2 + b^2 = (x_1 + x_2)^2 + (1 - x_1 x_2)^2 = (x_1^2 + 1)(x_2^2 + 1)$ must be a composite number.

> **Example 3.14**
>
> Given that the equation $x^2 - 2x - m = 0$ has no real roots. How many real roots
> does the equation $x^2 + 2mx + 1 + 2(m^2 - 1)(x^2 + 1) = 0$ have?
> **Solution**: None. Derive $m < -1$ from $\Delta_1 < 0$, and $\Delta_2 = -4(2m - 1)(m + 1)(2m + 1)(2m - 1) < 0$. Or show the left hand side is always positive.

> **Example 3.15**
>
> Let x_1, x_2 be the two roots for equation $x^2 + x - 3 = 0$, find the value of $x_1^3 - 4x_2^2 + 19$.
> **Solution**: Since $x_1^2 = 3 - x_1, x_2^2 = 3 - x_2$, we have $x_1^3 - 4x_2^2 + 19 = x_1(3 - x_1) - 4(3 - x_2) + 19 = 3x_1 - x_1^2 + 4x_2 + 7 = 3x_1 - (3 - x_1) + 4x_2 + 7 = 4(x_1 + x_2) + 4 = 0$.

Unless otherwise indicated, the unknown variable is x in the equations. Other variables such as m, k, p, q, etc., are parameters.

Problem 3.8 For the equation $x^2 - 402x + k = 0$, one of the roots plus three equals 80 times the other root. Find the value of k.

Problem 3.9 Without solving the equation, find the number of real roots for x: $(n^2 + 1)x^2 - 2nx + (n^2 + 4) = 0$.

Problem 3.10 For what values of m the equation $4x^2 + 8x + m = 0$ has two distinct real roots?

Problem 3.11 A quadratic equation has two roots $\frac{2}{3}$ and $-\frac{1}{2}$, what is this equation? (multiple answers are possible)

Problem 3.12 Two real numbers have sum -10 and product -5, find these two numbers.

Problem 3.13 Let a, b, c be real numbers. The following three quadratic equations each has only one real root (i.e. double roots):

$$\begin{aligned}
ax^2 + 2bx + c &= 0 \\
bx^2 + 2cx + a &= 0 \\
cx^2 + 2ax + b &= 0
\end{aligned}$$

Show that $a = b = c$.

Problem 3.14 Given that $x^2 + 2px + 1 = 0$ has two real roots, one is greater than 1, and the other is less than 1. What is the range of the possible values of p?

Problem 3.15 For equation $2x^2 + mx - 2m + 1 = 0$, the sum of squares of the two real roots is $\dfrac{29}{4}$. Find the value of m.

Problem 3.16 Let x_1 and x_2 be the two roots of the equation $4x^2 - 8x + k = 0$. Given that $\dfrac{1}{x_1} + \dfrac{1}{x_2} = \dfrac{8}{3}$, find k.

Problem 3.17 If x_1 and x_2 are the two real roots of $x^2 + x + q = 0$, and $|x_1 - x_2| = q$, find the value of q.

Problem 3.18 The equation $x^2 + (a-6)x + a = 0$ has two integer roots. Find the value of a.

Problem 3.19 Given equation in x: $x^2 + 2mx + m + 2 = 0$.

  (a) For what values of m does the equation have two positive roots?

  (b) For what values of m does the equation have one positive root and one negative root?

Problem 3.20 The quadratic equation $x^2 + 2kx + 2k^2 - 1 = 0$ has at least one negative root. Find the possible range of values for k.

Problem 3.21 Find the real solution to the system of equations: $x + y = 2$ and $xy - z^2 = 1$.

Problem 3.22 Given $p > 0$ and $q < 0$, how many positive roots does the equation $x^2 + px + q = 0$ have?

Problem 3.23 The sum of squares of the roots of equation $x^2 + 2kx = 3$ is 10. Find the possible values of k.

Problem 3.24 If x_1 and x_2 are integer roots of equation $x^2 + mx + 2 - n = 0$, and $(x_1^2 + 1)(x_2^2 + 1) = 10$, how many possible pairs (m, n) are there?

Problem 3.25 For the equation $x^2 + mx + n = 0$, the difference between the two roots is p and the product of the two roots is q. What is $m^2 + n^2$ in terms of p and q?

Problem 3.26 Let x_1, x_2 be two positive integer roots of equation $x^2 + px + 1997 = 0$.

Find the value of $\dfrac{p}{(x_1+1)(x_2+1)}$.

Problem 3.27 The two real roots of $x^2+(m-2)x+5-m=0$ are both greater than 2. Find the possible range of values for real number m.

Problem 3.28 Let m be integer. The equation $x^2+mx-m+1=0$ has two distinct positive integer roots. Find m.

3.2.2 Equations with Fractions

When unknowns appear on denominators, the standard method is to multiply everything by the denominator (or the least common multiple of the denominators) to get rid of the denominators. It is very likely to introduce extraneous roots. Therefore it is always necessary to verify the roots at the end. In some special cases, "change of variable" may be used to simplify the equations.

Example 3.16

$$\frac{4x}{x^2-4} - \frac{2}{x-2} = \frac{x+1}{x+2}.$$
Solution: Multiplying $(x+2)(x-2)$ to get $4x - 2(x+2) = (x+1)(x-2)$, solve and get roots 1 and 2. 2 is extraneous, so $x = 1$.

Example 3.17

$$\frac{1}{2x^2-3} - 8x^2 + 12 = 0.$$

Solution: Let $y = 2x^2 - 3$, then $\frac{1}{y} - 4y = 0$. Solve and get $y = \pm\frac{1}{2}$, and solve for x, get $x = \pm\frac{\sqrt{7}}{2}$ and $x = \pm\frac{\sqrt{5}}{2}$, all are verified to be solutions.

Example 3.18

The equation $\frac{x}{x-2} + \frac{x-2}{x} + \frac{2x-a}{x(x-2)} = 0$ has exactly one real root for x. Find all possible values of a, and the corresponding roots x.
Solution: $a = 7/2, x = 1/2; a = 8, x = -1; a = 4, x = 1$. Case 1: Set $\Delta = 0$ and solve for a. Case 2: if $x = 2$ is an extraneous root, then $a = 8$, and the valid root is $x = -1$. Case 3: if $x = 0$ is an extraneous root, then $a = 4$, and the valid root is $x = 1$.

Example 3.19

Solve: $x^2 + x + \dfrac{1}{x} + \dfrac{1}{x^2} = 0$. **Hint:** This is so-called "reciprocal equation", whose coefficients are symmetric. Use the change of variable: $y = x + \dfrac{1}{x}$.
Solution: $x = -1$.
Solve for $y = -2, y = 1$. These leads to $x = -1$ and no solutions.

3.2.3 Equations with Absolute Values

The absolute value of a real number a is defined as:

$$|a| = \begin{cases} a, & \text{if} \quad a \geq 0; \\ -a, & \text{if} \quad a < 0. \end{cases}$$

The standard method to deal with absolute values is case analysis: solve in intervals where the expressions inside the absolute value do not change signs. Sometimes the following techniques also help: (1) Change of variables; (2) Use the property that absolute values are always nonnegative.

Example 3.20

Solve: $|x - |2x + 1|| = 3$.
Solution: Case work: $x \geq -\dfrac{1}{2}$ or $x < -\dfrac{1}{2}$. If $x \geq -\dfrac{1}{2}$, $|x + 1| = 3$, so $x = 2$ or $x = -4$ (throw away -4). If $x < -\dfrac{1}{2}$, $|3x + 1| = 3$, then $x = \dfrac{2}{3}$ (throw away) or $x = -\dfrac{4}{3}$. Final solution: $x = 2$ or $x = -\dfrac{4}{3}$.

Example 3.21

Solve: $|x^2 - 11x + 10| = |2x^2 + x - 45|$.
Solution: It is a bit tedious to do case work on $x < 1$, $1 \leq x < 9/2$, $9/2 \leq x < 5$, $5 \leq x < 10$ and $x \geq 10$. The faster way is to solve $x^2 - 11x + 10 = 2x^2 + x - 45$ and $x^2 - 11x + 10 = -(2x^2 + x - 45)$, and get $x = -6 \pm \sqrt{91}$ and $x = \dfrac{5 \pm \sqrt{130}}{3}$.

Example 3.22

If $|m - 2009| = -(n - 2010)^2$, what is $(m-n)^{2011}$?
Solution: No squares are negative, and no absolute values are negative. So ♣
$m = 2009, n = 2010$, and $m - n = -1$. So $(m-n)^{2011} = -1$.

3.2.4 Equations with Radicals

If unknown variables appear inside radicals, the common method is to square (or cube, etc., depending on the order of the roots) both sides to remove the radicals. Sometimes the following methods also help: (1) Change of variables; (2) $\sqrt{a}$ is always nonnegative for $a \geq 0$.

Example 3.23

$3 - \sqrt{2x - 3} = x$.
Solution: $3 - x = \sqrt{2x - 3}$, so $9 - 6x + x^2 = 2x - 3$, and $x^2 - 8x + 12 = 0$, ♣
then $x = 2$ and $x = 6$. Check and find that $x = 2$ is the solution.

Example 3.24

$\sqrt{x + 3} - \sqrt{3x - 2} = -1$.
Solution: $\sqrt{x + 3} = \sqrt{3x + 2} - 1$, squaring, $x + 3 = 3x - 2 - 2\sqrt{3x - 2} + 1$, ♣
thus $x - 2 = \sqrt{3x - 2}$, square again, $x^2 - 7x + 6 = 0$, and get $x = 1$ or 6. $x = 1$
is extraneous. Therefore $x = 6$.

Problem 3.29 Solve the equation $\dfrac{3 - x}{2 + x} = 5 - \dfrac{4(2 + x)}{3 - x}$.

Problem 3.30 Solve: $\dfrac{15}{x + 1} = \dfrac{15}{x} - \dfrac{1}{2}$.

Problem 3.31 Solve: $\dfrac{x-3}{x+1} - \dfrac{x+1}{3-x} = \dfrac{5}{2}$.

Problem 3.32 Solve: $\left(\dfrac{x+1}{x^2-1}\right)^2 - 4\left(\dfrac{x+1}{x^2-1}\right) + 3 = 0$.

Problem 3.33 Solve: $\dfrac{3x-1}{x^2+1} - \dfrac{3x^2+3}{3x-1} = 2$.

Problem 3.34 Solve: $2x^4 - 9x^3 + 14x^2 - 9x + 2 = 0$.

Problem 3.35 Find all solutions to $|||x+1|-1|-1| = 1$.

Problem 3.36 Solve: $x^2 - \sqrt{3x^2 + 7} = 1$.

Problem 3.37 Solve: $2x^2 - \sqrt{4x^2 - 12x} = 6x + 4$.

Problem 3.38 Solve: $\sqrt{x^2 + 3x + 7} - \sqrt{x^2 + 3x - 9} = 2$.

Problem 3.39 Solve for x:

$$x^2 - \sqrt{x^2 - 3x + 5} = 3x + 1.$$

Problem 3.40 Let p, q be positive integers, and the roots of $px^2 - qx + 1985 = 0$ are both prime numbers. What is the value of $12p^2 + q$?

Problem 3.41 Let A, B, p be integers, and the two roots of $x^2 + px + 19 = 0$ are both exactly 1 greater than the two roots of $x^2 + Ax + B = 0$, respectively. Find the value of $B - A$.

Problem 3.42 If a, b are integers, and equation $ax^2 + bx + 1 = 0$ has two distinct positive roots, both less than 1. What is the minimum possible value of a?

Problem 3.43 The equation $2kx^2 + (8k + 1)x + 8k = 0$ has two distinct real roots for x. Find the range of values for k.

Problem 3.44 For what values of a, b the equation $x^2 + 2(1 + a)x + (3a^2 + 4ab + 4b^2 + 2) = 0$ has real roots?

Problem 3.45 The equation $|x^2 - 5x| = a$ has exactly two distinct real roots. What is the possible range of values for a?

Problem 3.46 Let a be a rational number. Suppose the roots of the equation $x^2 + 3(a-1)x + (2a^2 + a + b) = 0$ are always rational numbers, no matter what rational values a takes. What is the value of b?

Problem 3.47 Given an equation in x: $x^2 + (m-2)x + \frac{1}{2}m - 3 = 0$.

(a) Show that no matter what real value m takes, the equation always has two distinct real roots.

(b) Let x_1, x_2 be the two real roots of the given equation, and assume they also satisfy $2x_1 + x_2 = m + 1$, find the value of m.

Problem 3.48 Given a quadratic equation in x: $x^2 - 2(m-2)x + m^2 = 0$. Does there exist a real number m such that this equation has two real roots and the sum of squares of these two roots equals 56? If so, find the value of m; if not, explain why.

Problem 3.49 Given that $x = \dfrac{1}{\sqrt{3}-2}, y = \dfrac{1}{\sqrt{3}+2}$, evaluate $\dfrac{x^2+xy+y^2}{x+y}$.

Problem 3.50 Given that $\dfrac{x^2}{x^2-2} = \dfrac{1}{1-\sqrt{2}-\sqrt{3}}$, evaluate

$$\dfrac{\dfrac{1}{1-x}-\dfrac{1}{1+x}}{\dfrac{x}{x^2-1}+x}.$$

Problem 3.51 Real numbers x and y satisfy $|2x-y+1|+2\sqrt{3x-2y+4}=0$. Find the value of

$$1-\dfrac{x-y}{x-2y} \div \dfrac{x^2-y^2}{x^2-4xy+4y^2}.$$

Problem 3.52 Solve the equation:

$$\sqrt{1 + \frac{2}{x-1}} - \sqrt{1 - \frac{2}{x+1}} = \frac{3}{2}.$$

Problem 3.53 Let a be a real number, and the equation $x^2 + a^2 x + a = 0$ has real roots for x. Find the maximum possible root x.

Problem 3.54 Solve for x: $(x - \sqrt{3})x(x+1) + 3 - \sqrt{3} = 0$.

Problem 3.55 Solve: $\sqrt{\sqrt{\sqrt{x+4}+4}} = x$

Problem 3.56 Solve for x: $\sqrt{5 - \sqrt{5-x}} = x$.

3.3   Factoring Polynomials Continued

In this section we explore some more advanced techniques for factoring.

3.3.1   Change of variables

Use a new variable to replace a complicated part of the original expression, to make a simpler new expression.

Example 3.25

Factor $(x^2+x+1)(x^2+x+2)-12$.
Solution: Let $y=x^2+x+1$, the expression becomes $y(y+1)-12=y^2+y-12=(y+4)(y-3)$, then change back to x.
Answer: $(x-1)(x+2)(x^2+x+5)$

Example 3.26

Factor $(x^2+3x+2)(x^2+7x+12)-120$.
Solution: $(x^2+3x+2)(x^2+7x+12)-120=(x+1)(x+2)(x+3)(x+4)-120=(x^2+5x+4)(x^2+5x+6)-120$. Let $y=x^2+5x+5$, then $(y-1)(y+1)-120=y^2-121=(y+11)(y-11)$.
Answer: $(x^2+5x+16)(x+6)(x-1)$.

Problem 3.57 Factor $(x^2+3x+2)(4x^2+8x+3)-90$.

Problem 3.58 Factor $(x^2+4x+8)^2+3x(x^2+4x+8)+2x^2$.

Problem 3.59 Factor $6x^4 + 7x^3 - 36x^2 - 7x + 6$.

Problem 3.60 Factor $(x^2 + xy + y^2)^2 - 4xy(x^2 + y^2)$.

Problem 3.61 Let n be an integer, show that $n(n+1)(n+2)(n+3) + 1$ is a perfect square.

3.3.2 Finding roots

Two important theorems are used for this technique.

> **Theorem 3.2 Factor Theorem**
>
> Let $P(x) = a_n x^n + a_{n-1} x^{n-1} + \cdots + a_1 x + a_0$ be a polynomial, and $a \in \mathbb{R}$ be a root of $P(x)$, that is, $P(a) = 0$, then $P(x)$ has a factor $(x-a)$.

> **Theorem 3.3 Rational Roots Theorem**
>
> Let $P(x) = a_n x^n + a_{n-1} x^{n-1} + \cdots + a_1 x + a_0$ be a polynomial with integer coefficients, and suppose $\dfrac{p}{q}$ is a rational root of $P(x)$, and $\gcd(p,q) = 1$, then $p \mid a_0$, $q \mid a_n$.

> **Example 3.27**
>
> Factor $x^3 - 4x^2 + 6x - 4$.
> **Solution**: Let $x = 2$, the value of the expression is 0, thus $x - 2$ is a factor.
> Answer: $(x-2)(x^2 - 2x + 2)$

> **Example 3.28**
>
> Factor $9x^4 - 3x^3 + 7x^2 - 3x - 2$.
> **Solution**: Possible roots have the form $\pm p/q$, where $p = 1$ or 2, and $q = 1, 3$ or 9. Try and verify that $-1/3$ and $2/3$ are roots. Answer: $(3x+1)(3x-2)(x^2+1)$

Problem 3.62 Factor $x^3 - 19x - 30$.

Problem 3.63 Show that $2x + 3$ is a factor of $2x^4 - 5x^3 - 10x^2 + 15x + 18$.

3.3.3 Method of undetermined coefficients

Sometimes we can predict the form of the resulting factorization, but some coefficients are not known, then we use variables to indicate those undetermined coefficients, and find them by (1) comparing coefficients; or (2) by plugging in numbers for the originally-given variables.

> **Example 3.29**
>
> Factor $x^2 + 3xy + 2y^2 + 4x + 5y + 3$.
> **Solution:** $(x+2y+3)(x+y+1)$. Factor $x^2 + 3xy + 2y^2 = (x+y)(x+2y)$, and ♣
> let $x^2 + 3xy + 2y^2 + 4x + 5y + 3 = (x+2y+m)(x+y+n)$.

> **Example 3.30**
>
> Factor $x^4 - 2x^3 - 27x^2 - 44x + 7$.
> **Solution:** $(x^2 - 7x + 1)(x^2 + 5x + 7)$. $x^4 - 2x^3 - 27x^2 - 44x + 7 = (x^2 + ax +$ ♣
> $b)(x^2 + cx + d)$, try the cases $b = 1, d = 7$ and $b = -1, d = -7$.

Problem 3.64 If k is an integer and $x^2 + 2kx - 3k^2$ has a factor $(x-1)$, what should k be?

Problem 3.65 $x^4 + 5x^3 + 15x - 9$

Problem 3.66 $x^4 - 12x + 323$

Problem 3.67 $x^3 + 3x^2 - 4$

Problem 3.68 $x^4 - 11x^2y^2 + y^4$

Problem 3.69 $x^3 + 9x^2 + 26x + 24$

Problem 3.70 $(2x^2 - 3x + 1)^2 - 22x^2 + 33x - 1$

Problem 3.71 $x^4 + 7x^3 + 14x^2 + 7x + 1$

Problem 3.72 $(x+3)(x^2-1)(x+5)-20$

Problem 3.73 $2x^2+3xy-9y^2+14x-3y+20$

Problem 3.74 $a^2+(a+1)^2+(a^2+a)^2$

Problem 3.75 $2acx+4bcx+adx+2bdx+4acy+8bcy+2ady+4bdy$

Problem 3.76 $ab(c^2-d^2)-cd(a^2-b^2)$

Problem 3.77 $a^5 + a + 1$.

Problem 3.78 $x^4 + y^4 + z^4 - 2x^2y^2 - 2y^2z^2 - 2z^2x^2$

Problem 3.79 $1 + 2a + 3a^2 + 4a^3 + 5a^4 + 6a^5 + 5a^6 + 4a^7 + 3a^8 + 2a^9 + a^{10}$.

Problem 3.80 $(x+1)(x+3)(x+5)(x+7) + 15$

Problem 3.81 Evaluate the following: $\dfrac{(1994^2 - 2000)(1994^2 + 3985) \times 1995}{1991 \cdot 1993 \cdot 1995 \cdot 1997}$.

3.4 Polynomials

3.4.1 Definitions related to polynomials

- **Polynomial.** A function that is made of adding multiples of powers of a variable. Examples: $5x^3 - 2x + 1$, $-a^{99} - 9a^9 - 99$, and $4z^2 - 12z + 9$. Usually polynomials are written with the powers going up or going down.

- **Monomial (or term).** Both words refer to a polynomial with exactly one piece. Every polynomial can be considered a sum of monomials / terms. For example, the polynomial $x^7 - 4x^5 + 18x$ is the sum of the monomials x^7, $-4x^5$, and $18x$.

- **Degree (of a polynomial).** The largest power in a polynomial. The degrees of the polynomials $5x^3 - 2x + 1$, $-a^{99} - 9a^9 - 99$, and $4z^2 - 12z + 9$ are 3, 99, and 2. A constant is also a polynomial, and it has degree 0 if the constant itself is nonzero.

- **The zero polynomial.** The function $p(x) = \mathbf{0}$ is a polynomial as well, called the "zero polynomial". Note that this is not an equation. The polynomial always has value 0 no matter what x is. The zero polynomial has no terms and, strictly speaking, it has no degree either. Sometimes it is convenient to define the degree of the zero polynomial to be either -1 or negative infinity ($-\infty$).

- **Coefficient (of a term).** The number being multiplied by a power. The coefficient of x^3 in $5x^3 - 2x + 1$ is 5; the coefficient of a^9 in $-a^{99} - 9a^9 - 99$ is -9; the coefficient of z^7 in $4z^2 - 12z + 9$ is 0.

- **Function notation.** Often it is convenient to use function notation to represent long polynomials. When we say $f(x) = 5x^3 - 2x + 1$, we are saying that f "assigns" to a number x the value $5x^3 - 2x + 1$.

- **Zero of a polynomial** (also called **"root"** of a polynomial equation). The value of x where the polynomial $P(x)$ has the value 0. It is a solution of the polynomial equation $P(x) = 0$.

- **Factored polynomial.** A polynomial is sometimes written as a product of different polynomials. For example, $(3x + 1)(2x - 7)$, or $(z^2 + 2z + 3)(-2z + 5)(7z^7 + 4z^4)$, or $(2y - 8)^2(5y^2)^3$. This is sometimes done to make equations easier to solve, or to make values easier to compute.

- **Polynomial of multiple variables.** There can be multiple variables in a polynomial: $3xy^2$, $9x^5y^5 + 20x^3y^4$. The **degree** of these polynomials is the maximum sum of the exponents of variables. For example, $3xy^2$ has degree 3, and $9x^5y^5 + 20x^3y^4$ has degree 10.

3.4.2 Facts about polynomials

The following are always true.

- If the degree of a polynomial $P(x)$ is d, then the number of terms of $P(x)$ is at least 1 and at most $d+1$.

> **Example 3.31**
>
> The polynomials $2a^3$, $3x^3 - 2x$, $7t^3 + 5t^2 + 2t$, and $100u^3 + 4u^2 - 3u - 20$ are all polynomials of degree 3. ♣

- If the degrees of polynomials $p(y)$ and $q(y)$ are d and e, then the degree of $p(y) \cdot q(y)$ is $d + e$.

> **Example 3.32**
>
> Let $p(y) = x^2 + 2$ (degree 2) and $q(y) = x^3 - x$ (degree 3), then $p(y)q(y) = x^5 + x^3 - 2x$ (degree 5). ♣

- If the degrees of polynomials $N(y)$ and $M(y)$ are d and e, then the degree of $N(M(y))$ is $d \cdot e$.

> **Example 3.33**
>
> Let $N(y) = x^2 + 2$ (degree 2) and $M(y) = x^3 - x$ (degree 3), then $N(M(y)) = (x^3 - x)^2 + 2 = x^6 - 2x^4 + x^2 + 2$ (degree 6). ♣

- If the degree of every term in the polynomial $g(x)$ is even, and every coefficient is positive, then $g(x) \geq 0$ for every possible real value of x.

> **Example 3.34**
>
> Consider the polynomial $x^4 + 100x^2 + 2$. It's value is nonnegative (in fact, always positive) for all real values of x. ♣

In this section, the solutions to the equations *do not have to be real numbers*, unless explicitly specified.

Problem 3.82 Distinguish the following concepts. Find examples for each of them.

(a) A zero-degree polynomial.

(b) The zero polynomial.

(c) Zero of a polynomial.

Problem 3.83 For this problem, $f(x) = 3x + 2$, $g(x) = x - 7$, and $h(x) = x^2 - 4x + 4$. Compute the following values:

(a) $f(g(4))$

(b) $g(f(4))$

(c) $g(g(g(g(g(35)))))$

(d) $h(f(0))$

(e) $h(f(1))$

(f) $h(f(100))$

(g) $f(g(1234567)) - g(f(1234567))$

Problem 3.84 In the polynomial $(7+x)(1+x^2)(5+x^4)(2+x^8)(3+x^{16})(10+x^{32})$, what is the coefficient of x^{54}?

3.4.3 Useful Theorems

Theorem 3.4 Fundamental Theorem of Algebra

(also known as the **Gauss-d'Alembert Theorem**) Every non-constant polynomial with complex coefficients has at least one complex zero. Consequently, the number of zeros of a polynomial equals the degree, multiplicities counted.

Note: The proof of this theorem is beyond the scope of this book. You can find more information of this theorem online.

Theorem 3.5 Polynomial Remainder Theorem

The remainder of a polynomial $P(x)$ divided by a linear divisor $x - a$ is equal to $P(a)$.

Proof. Consider the polynomial division formula: Given polynomials $P(x)$ and $D(x)$, we can always write

$$P(x) = D(x)Q(x) + R(x),$$

where $\deg R(x) < \deg D(x)$. Here $Q(x)$ is the quotient, and $R(x)$ is the remainder. In the case of a linear divisor $D(x) = x - a$, the remainder is a constant r. So

$$P(x) = (x - a)Q(x) + r.$$

In the above identity, let $x = a$, then $P(a) = r$. ∎

Theorem 3.6 Factor Theorem

A polynomial $P(x)$ has a factor $x - a$ if and only if $P(a) = 0$.

Proof. This is just a special case (also called a "corollary") of the Polynomial Remainder Theorem when $r = 0$. ∎

Theorem 3.7 Vieta's Theorem (quadratic version)

Let x_1, x_2 be the roots of a quadratic equation $ax^2 + bx + c = 0$, where $a \neq 0$, then

$$x_1 + x_2 = -\frac{b}{a}, \quad \text{and} \quad x_1 x_2 = \frac{c}{a}.$$

Proof. This is easily proven using the quadratic formula. Also see the proof of the general form below.   ∎

> ### Theorem 3.8 Vieta's Theorem (general version)
>
> Let $P(x) = a_n x^n + a_{n-1} x^{n-1} + \cdots + a_0$ be a polynomial of degree n, and $x_1, x_2, \ldots, x_n$ be the zeros of $P(x)$. Then
>
> $$\begin{cases} x_1 + x_2 + \cdots + x_n = -\dfrac{a_{n-1}}{a_n}, \\[2mm] x_1 x_2 + x_1 x_3 + \cdots + x_{n-1} x_n = \dfrac{a_{n-2}}{a_n}, \\[2mm] \cdots \\[2mm] x_1 x_2 \cdots x_n = (-1)^n \dfrac{a_0}{a_n}. \end{cases}$$

Proof. By the Factor Theorem, since $x_1, x_2, \ldots, x_n$ are the zeros of $P(x)$, each of the linear polynomials $(x - x_1), (x - x_2), \ldots, (x - x_n)$ is a factor of $P(x)$. Thus

$$P(x) = a_n (x - x_1)(x - x_2) \cdots (x - x_n).$$

Expanding the right hand side, and compare the coefficients with

$$P(x) = a_n x^n + a_{n-1} x^{n-1} + \cdots + a_0$$

to get the desired identities.   ∎

Problem 3.85 The two roots of equation $x^2 + px + 1 = 0 (p > 0)$ has a difference 1. Find the value of p. .

Problem 3.86 Let $m \geq -1$ be a real number, and the equation $x^2 + 2(m - 2)x + m^2 - 3m + 3 = 0$ has two distinct real roots x_1 and x_2. If $x_1^2 + x_2^2 = 6$, what is m?

Problem 3.87 Expand $(x^2 - x + 1)^6$ to get $a_{12}x^{12} + a_{11}x^{11} + \cdots + a_1x + a_0$. Find the value of $a_{12} + a_{10} + a_8 + a_6 + a_4 + a_2 + a_0$.

Problem 3.88 Assume $(x - c)^2 \mid (4x^3 + 8x^2 - 11x + 3)$, find the value of c.

Problem 3.89 Assume $(x - 1)^2 \mid [x^4 + (m+n)x^3 + (m-n)x^2 + (m^2 + 2n - 1)x + m + 2]$. Find the value of m and n.

Problem 3.90 Let $x = \dfrac{2}{2 + \sqrt{3} - \sqrt{5}}, y = \dfrac{2}{2 + \sqrt{3} + \sqrt{5}}$, evaluate:

$$\frac{x^4y^4}{x^4 + y^4 + 6x^2y^2 + 4x^3y + 4xy^3}.$$

Problem 3.91 Distinct real numbers a and b satisfies $(a+1)^2 = 3 - 3(a+1)$, $3(b+1) = 3 - (b+1)^2$. Find the value of $b\sqrt{\dfrac{b}{a}} + a\sqrt{\dfrac{a}{b}}$.

Problem 3.92 Without solving the equation, find out the number of real roots of the following equation: $x^3 + 3x - 1 = 0$.

Problem 3.93 Let m, p be positive integers. The two parabolas $P_1(x) = x^2 + 5x + m$ and $P_2(x) = x^2 + px + 2$ have no common points, and $P_1(100) < P_2(100)$. Find the value of $m + p$.

Problem 3.94 An $l \times w \times h$ rectangular box has surface area 38 and volume 12. If $l + w + h = 8$, find the dimensions of the box.

Problem 3.95 Let a, b, c, and d be the roots of $x^4 - 2x - 1990 = 0$. Find the value of $1/a + 1/b + 1/c + 1/d$.

Problem 3.96 If a, b, c, d are four different numbers for which

$$\begin{cases} a^4 + a^2 + ka + 64 = 0 \\ b^4 + b^2 + kb + 64 = 0 \\ c^4 + c^2 + kc + 64 = 0 \\ d^4 + d^2 + kd + 64 = 0. \end{cases}$$

What is the value of $a^2 + b^2 + c^2 + d^2$?

Problem 3.97 Find the sum of the 17th powers of the 17 roots of $x^{17} - 3x + 1 = 0$.

Problem 3.98 Let x and y be nonzero real numbers satisfying $|x| + y = 3$ and $|x|y + x^3 = 0$, Find the value of $x + y$.

Problem 3.99 Find ordered pairs (x,y) of real numbers such that $x^2 - xy + y^2 = 13$ and $x - xy + y = -5$.

Problem 3.100 If $x + y + z = 0$ and $x^3 + y^3 + z^3 = 288$, find the value of xyz.

Problem 3.101 Let x be a real number such that $x^3 + 4x = 8$. Determine the value of $x^7 + 64x^2$.

Problem 3.102 The polynomial $p(x) = x^3 + 2x^2 - 5x + 1$ has three different roots a, b, and c. Find $a^3 + b^3 + c^3$.

Problem 3.103 Suppose that the roots of $3x^3 + 3x^2 + 4x - 11 = 0$ are a, b and c, and the roots of $x^3 + rx^2 + sx + t = 0$ are $a + b, b + c$, and $c + a$. Find t.

Problem 3.104 Let $P(x) = (x-1)(x-2)(x-3)$. For how many polynomials $Q(x)$ does there exist a polynomial $R(x)$ of degree 3 such that $P(Q(x)) = P(x)R(x)$?

Problem 3.105 Determine $x^2 + y^2 + z^2 + w^2$ if

$$\frac{x^2}{2^2 - 1^2} + \frac{y^2}{2^2 - 3^2} + \frac{z^2}{2^2 - 5^2} + \frac{w^2}{2^2 - 7^2} = 1,$$

$$\frac{x^2}{4^2 - 1^2} + \frac{y^2}{4^2 - 3^2} + \frac{z^2}{4^2 - 5^2} + \frac{w^2}{4^2 - 7^2} = 1,$$

$$\frac{x^2}{6^2 - 1^2} + \frac{y^2}{6^2 - 3^2} + \frac{z^2}{6^2 - 5^2} + \frac{w^2}{6^2 - 7^2} = 1,$$

$$\frac{x^2}{8^2 - 1^2} + \frac{y^2}{8^2 - 3^2} + \frac{z^2}{8^2 - 5^2} + \frac{w^2}{8^2 - 7^2} = 1.$$

4. Geometry

Geometry is an ancient subject because it was one of the first practical applications of mathematics. As people studied different shapes and forms they discovered there was structure attached to the forms. In this chapter we shall explore the various shapes such as lines, angles, polygons, circles, and measurements like lengths, angle measures, areas, etc.

4.1 Angles

Angles Review

- Angle: an *angle* is formed when two rays have a common origin. The common origin is the *vertex* of the angle, and the two rays are the *sides*.
- An entire circle is 360 degrees, or 360°.
- A *right angle* is 90°. Two lines (or rays or segments) that form a right angle are *perpendicular*. If $\angle BAC = 90°$, then $\overline{BA}$ and $\overline{AC}$ are perpendicular, or $\overline{BA} \perp \overline{AC}$.
- An angle less than 90° is called an *acute angle*.
- An angle greater than 90° but less than 180° is called an *obtuse angle*.
- A 180° angle is in fact a straight line, so it is called a *straight angle*.
- An angle greater than 180° is called a *reflex* angle.
- Two angles that add up to 180° are called *supplementary angles*.
- Two angles that add up to 90° are called *complementary angles*.

Parallel Lines

- Let *m* and *n* be a pair of parallel lines. A third line *l* cuts across the parallel lines. Line *l* is called a *transversal*.

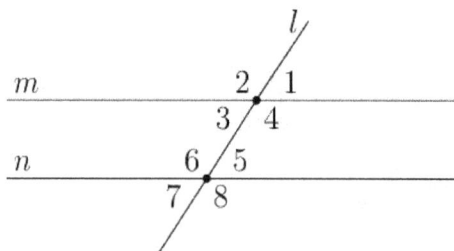

Figure 4.1: Parallel lines, transversal, and related angles

- In Figure 4.1,
    - *Corresponding angles*: $\angle 1 = \angle 5$
    - *Alternate interior angles*: $\angle 3 = \angle 5$
    - *Alternate exterior angles*: $\angle 2 = \angle 8$
    - *Same-side interior angles*: $\angle 4 + \angle 5 = 180°$
    - *Same-side exterior angles*: $\angle 1 + \angle 8 = 180°$
  In summary: $\angle 1 = \angle 3 = \angle 5 = \angle 7$, and $\angle 2 = \angle 4 = \angle 6 = \angle 8$. Also, $\angle 4 + \angle 5 = \angle 3 + \angle 6 = \angle 1 + \angle 8 = \angle 2 + \angle 7 = 180°$.

Special Angles

- Special angles are the angles $30°$, $60°$, $90°$ and $45°$.
- An equilateral triangle has three $60°$ angles.
- As shown in Figure 4.2, in $\triangle ABC$, if $\angle C = 90°$, $\angle A = 60°$, and $\angle B = 30°$, then $AB = 2AC$, $BC = \sqrt{3}AC$.

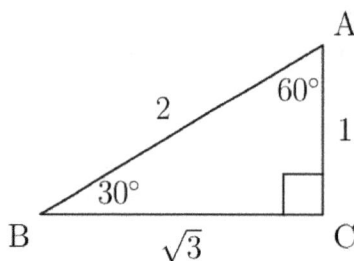

Figure 4.2: A 30-60-90 Triangle

- **Notation:** we use $[ABC]$ to denote the area of triangle ABC, $[DEFG]$ to denote the area of quadrilateral $DEFG$, etc.

Problem Solving Strategies

- Draw accurate diagrams using compass, ruler, and protractor.
- Try to find or make special angles in the diagram.

Problem 4.1 An equilateral triangle must also be equiangular, and vice versa. But this is not true for other polygons.

   (a) Give an example of an equiangular polygon that is not equilateral.

   (b) Give an example of an equilateral polygon that is not equiangular.

Problem 4.2 Complete the following table about polygons: name, sum of interior angles, sum of exterior angles, and measure of each angle in case of regular polygon. All angles are in degrees. Justify your answers.

# sides	Polygon	Int. angle sum	Ext. angle sum	Each angle (if regular)
3	Triangle			
4				
5				
6				
7	Heptagon			
8				
9	Nonagon			
10				
12	Dodecagon			
20	Icosagon			

Problem 4.3 Prove the following: in a triangle, an exterior angle equals the sum of the two interior angles not adjacent to it.

Problem 4.4 In a regular hexagon $ABCDEF$:

(a) Is ACE an equilateral triangle? Justify your answer.

(b) Calculate the angle AED.

Problem 4.5 Suppose that $ABCD$ is a square. Let point E be *outside* the square and that $\triangle CDE$ is an equilateral triangle (see the diagram). What is the measure of $\angle EAD$?

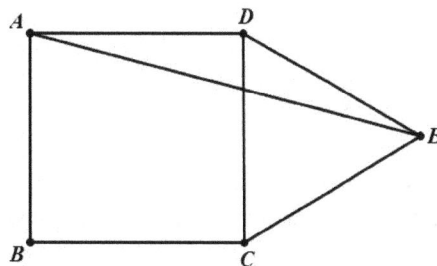

Problem 4.6 Same question as above, but with *E inside* the square.

Problem 4.7 Given square *ABCD*, let *P* and *Q* be the points outside the square that make triangles *CDP* and *BCQ* equilateral. Segments $\overline{AQ}$ and $\overline{BP}$ intersect at *G*. Find angle *AGP*.

Problem 4.8 Mark *P* inside square *ABCD*, so that triangle *ABP* is equilateral. Let *Q* be the intersection of *BP* with diagonal *AC*. Triangle *CPQ* looks isosceles. Is this actually true?

Problem 4.9 Point *P* is inside regular pentagon *ABCDE* so that triangle *ABP* is equilateral. Decide whether or not quadrilateral *ABCP* is a parallelogram, and give your reasons.

Problem 4.10 In $\triangle ABC$, $\angle C = 2\angle A$, $AC = 2BC$. Find the measure of $\angle B$.

Problem 4.11 In $\triangle ABC$, $\angle ABC = 12°$, $\angle ACB = 132°$. Let $\overline{BM}$ and $\overline{CN}$ be the exterior angle bisectors where M and N are on the lines $\overline{AC}$ and $\overline{AB}$ respectively. Then
(A) $BM > CN$ (B) $BM = CN$ (C) $BM < CN$ (D) Can't determine.

Problem 4.12 Given square $ABCD$, let P and Q be the points outside the square that make triangles CDP and BCQ equilateral. Prove that triangle APQ is also equilateral. What if $ABCD$ is a rectangle? What if $ABCD$ is a parallelogram?

Problem 4.13 Equilateral triangles BCP and CDQ are attached to the outside of regular pentagon $ABCDE$. Is quadrilateral $BPQD$ a parallelogram? Justify your answer.

Problem 4.14 Three non-overlapping regular plane polygons all have sides of length 1.

The polygons meet at a point A in such a way that the sum of the three interior angles at A is $360°$. Thus the three polygons form a new polygon P (not necessarily convex) with A as an interior point. Among the three polygons, one is a square and one is a pentagon. Find the perimeter of P.

Problem 4.15 Find the side of the largest square that can be drawn inside an equilateral triangle with side length 12, one side of the square aligned with one side of the triangle.

Problem 4.16 The equiangular convex hexagon $ABCDEF$ has $AB = 1, BC = 4, CD = 2$, and $DE = 4$. Find $[ABCDEF]$.

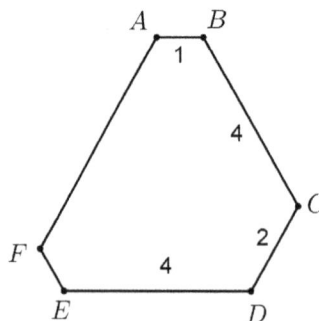

Problem 4.17 A stop sign — a regular *octagon* — can be formed from a 12-inch square sheet of metal by making four straight cuts that snip off the corners. How long are the sides of the resulting polygon?

Problem 4.18 The Golden Triangle. In $\triangle ABC$, $AB = AC$. Point D is on side $\overline{AB}$ such that $\overline{CD}$ bisects $\angle ACB$, and $CD = BC$. Find the angles of $\triangle ABC$.

Problem 4.19 $BCDE$ is a square, $\triangle ABC \cong \triangle FCD$ with $\angle A = 120°$ and $AB = AC$. If $AF = 20$, compute the area of square $BCDE$.

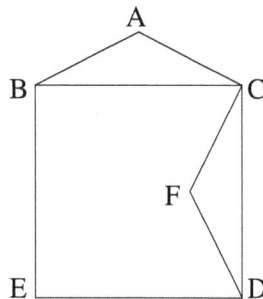

Problem 4.20 In $\triangle ABC$, $\angle B = 90°$ and $\angle C = 30°$. Point D is on $\overline{BC}$ such that $\angle ADB =$

45°, and $DC = 10$. What is AB?

Problem 4.21 Let $ABCD$ be a square. Through A construct line $\overline{AP}$ outside the square so that $\angle DAP = 45°$. Let E be a point on $\overline{AP}$ so that $BE = BD$. Let F be the intersection of $\overline{BE}$ and $\overline{AD}$. Is $\triangle DEF$ isosceles? Justify your answer.

Problem 4.22 Let ABC be an isosceles right triangle with $\angle C = 90°$. Let P be a point inside the triangle such that $AP = 3, BP = 5$, and $CP = 2\sqrt{2}$. What is $[ABC]$?

4.2 Area Methods

4.2.1 Fundamentals

Notation: We shall use $[ABC]$ to denote the area of triangle ABC, $[XYZW]$ to denote the area of the quadrilateral $XYZW$, etc.

Formulas for areas of square, rectangle, triangle, parallelogram, trapezoid, circle, etc. should be memorized.

Various area formulas of triangle ABC:

$$[ABC] = \frac{1}{2}ah \qquad (h \text{ is the altitude on } a)$$

$$= \frac{abc}{4R} \qquad (R \text{ is the circumradius})$$

$$= \sqrt{s(s-a)(s-b)(s-c)} \qquad \left(s = \frac{a+b+c}{2}\right)$$

$$= rs \qquad (r \text{ is the inradius, } s \text{ is defined as above})$$

Simple (but very useful) Facts:

- The areas of triangles (or parallelograms) with equal bases and equal altitudes (heights) are equal.
- The areas of triangles with equal altitudes are proportional to the bases of the triangles.
- The ratio of areas between two similar triangles is the square of the ratio between the corresponding sides.

Problem 4.23 Prove the Pythagorean Theorem using areas.

Problem 4.24 In $\triangle ABC$, $AB > AC > BC$, $\overline{CD}, \overline{BE}, \overline{AF}$ are altitudes on $\overline{AB}, \overline{AC}, \overline{BC}$, respectively. Show that $CD < BE < AF$.

Problem 4.25 In triangle ABC, $AC = 10$, $BC = 24$, $AB = 26$. What is the altitude on $\overline{AB}$?

Problem 4.26 Let $ABCD$ be a parallelogram, and E, H, F, G be points on sides $\overline{AB}$, $\overline{BC}$, $\overline{CD}$, $\overline{DA}$ respectively, and $\overline{EF} \| \overline{BC}$ and $\overline{GH} \| \overline{AB}$. Let P be the intersection of $\overline{EF}$ and $\overline{GH}$. If $[GPFD] = 10, [PHCF] = 8, [EBHP] = 16$, find $[ABCD]$.

Problem 4.27 In $\triangle ABC$, let D, E, F be midpoints of the sides $\overline{BC}, \overline{AC}, \overline{AB}$. Show that $[DEF] = [ABC]/4$.

Problem 4.28 Suppose you only know that the centroid exists. Prove (using areas!) that the centroid divides each median in a ratio of $1 : 2$.

Problem 4.29 Let $ABCD$ be a parallelogram with $[ABCD] = 1$. Let P be a point in the interior of $ABCD$. Show that $[ABP] + [CDP]$ is a fixed value and find that value.

Problem 4.30 Let $ABCD$ be a parallelogram, with midpoints E, F, G, H (say on $\overline{AB}, \overline{BC}, \overline{CD}, \overline{DA}$). Let I, J be the midpoints of $\overline{EF}, \overline{GH}$. Find the area of $\triangle JIG$ as a fraction of the area of $ABCD$.

Problem 4.31 Prove that if a triangle has side lengths a, b, c, inradius r, and circumradius R we have $2Rr = \dfrac{abc}{a+b+c}$.

Problem 4.32 Suppose the altitudes of a triangle are in ratio $2 : 2 : 3$ and the triangle has a perimeter of 24. Find the area of the triangle.

Problem 4.33 Given square $ABCD$, let E, F be the midpoints of AB and BC respectively, and G be the intersection of AF and CE. If $[ABCD] = 1$, find $[AGCD]$.

Problem 4.34 Suppose you have a trapezoid $ABCD$ with $\overline{AB}$ parallel to $\overline{CD}$. Let E be the intersection of the diagonals. Suppose $AB = 10, CD = 15$ and $\triangle ADE$ has area 24. Find the area of $ABCD$.

Problem 4.35 Prove the converse of the Pythagorean Theorem. Hint: You can use the Pythagorean Theorem!

Problem 4.36 Suppose you have a circle with diameter $\overline{AB}$ with $AB = 4$. Let C, D be on arc $\widehat{AB}$ such that $\widehat{AC} : \widehat{CD} : \widehat{DB} = 1 : 2 : 1$. Find the area of the figure enclosed by line segment $\overline{AC}$, arc $\widehat{CD}$, and line segment $\overline{AD}$.

Problem 4.37 In triangle ABC, $AC = 9$, $BC = 10$, $AB = 17$. What is the altitude on

$\overline{BC}$?

Problem 4.38 Let $ABCD$ be a rectangle, and E, H, F, G be points on sides $\overline{AB}$, $\overline{BC}$, $\overline{CD}$, $\overline{DA}$ respectively, and $\overline{EF} \parallel \overline{BC}$ and $\overline{GH} \parallel \overline{AB}$. Let P be the intersection of EF and GH. If $[GPFD] = 10, [EBHP] = 12$ and $[ABCD] = 44$. Suppose each of the four rectangles formed above has integer dimensions, find all possible dimensions for $GPFD$ and $EBHP$.

Problem 4.39 Find a formula for the area of a parallelogram if you are given the two sides lengths as well as one of the diagonals.

Problem 4.40 Prove that if a triangle has side lengths a, b, c semiperimeter s, inradius r, and circumradius R we have

$$\frac{R}{r} = \frac{abc}{4(s-a)(s-b)(s-c)}.$$

Problem 4.41 Suppose a parallelogram P_1 has area 256. Connect the midpoints of each side to form a parallelogram P_2. Repeat to get P_3, and continue repeating until you get to P_{10}.

(a) Prove this problem actually makes sense. That is, prove that if you connect the midpoints of a parallelogram you get another parallelogram.

(b) Find the area of P_{10}.

Problem 4.42 Let $ABCD$ be a trapezoid with $\overline{AB} \| \overline{CD}$. Let E be the intersection of the two diagonals $\overline{AC}, \overline{BD}$.

(a) If $ABCD$ is a parallelogram show $[ABE] = [BCE] = [CDE] = [DAE]$.

(b) Prove that for a general trapezoid, $[BCE] = [DAE]$.

Problem 4.43 Let $\overline{AM}$ be a median of $\triangle ABC$, and D be a point on $\overline{MC}$, and E be a point on $\overline{AB}$, such that $\overline{ME} \| \overline{AD}$. Show that $[BDE] = [AEDC]$.

4.2.2 Two Theorems Using Areas

Recall that in a triangle, a line segment connecting a vertex with a point on the opposite side is called a *cevian*.

Theorem 4.1 Ceva's Theorem

In triangle ABC, if three cevians AX, BY, CZ are concurrent, then

$$\frac{BX}{XC} \cdot \frac{CY}{YA} \cdot \frac{AZ}{ZB} = 1.$$

Proof. Use areas (see Problem 4.44).   ■

Remark

"Ceva" above is pronounced "chayva". The theorem was proved by Italian mathematician Giovanni Ceva.

Theorem 4.2 Converse of Ceva's Theorem

In triangle ABC, if three cevians AX, BY, CZ satisfy

$$\frac{BX}{XC} \cdot \frac{CY}{YA} \cdot \frac{AZ}{ZB} = 1,$$

then they are concurrent.

Proof. Use an indirect proof (see Problem 4.45).   ■

Theorem 4.3 Angle Bisector Theorem

In triangle ABC, let D be the point on $\overline{BC}$ such that $\overline{AD}$ bisects $\angle BAC$. Then $AB/AC = BD/DC$.

Proof. There are many different proofs using similar triangles. Try using an area argument (see Problem 4.46).   ■

Problem 4.44 Prove Ceva's Theorem using areas.

Problem 4.45 Prove the converse to Ceva's Theorem using an indirect proof.

Problem 4.46 Prove the Angle Bisector Theorem using areas.

Problem 4.47 **Centers of Triangles**: Prove the following, using Ceva's Theorem (or its converse) if applicable.

(a) (Centroid). In every triangle ABC, the three medians (i.e. the line from a vertex to the midpoint of the opposite side) are concurrent. This point is called the *centroid* of the triangle ABC.

(b) (Incenter). In every triangle ABC, the three angle bisectors (i.e. the line from a vertex bisecting the angle at that vertex) are concurrent. This point is called the *incenter* of the triangle ABC, because it is the center of the circle inscribed in

4.2 Area Methods

(c) (Orthocenter). In every triangle ABC, the three altitudes (i.e. the line from a vertex perpendicular to the opposite side) are concurrent. The common point of intersection is called the *orthocenter* of triangle ABC.

(d) (Circumcenter). In every triangle ABC, the perpendicular bisectors of the three sides are concurrent. This point is called the *circumcenter* of the triangle ABC, because it is the center of the circle circumscribed around triangle ABC.

Problem 4.48 In $\triangle ABC$, $AB = AC$, and P is a point on the side $\overline{BC}$. Prove that the sum of the distances from P to the sides $\overline{AB}$ and $\overline{AC}$ is a fixed value.

Problem 4.49 Let G be the centroid of $\triangle ABC$, and $AG = 3, BG = 4, CG = 5$, find $[ABC]$.

(blank)
(blank)

Copyright © Areteem Institute. All rights reserved.

Problem 4.50 Let ABC be a triangle with ω as its circumcircle. Arcs $\overset{\frown}{AB}$, $\overset{\frown}{BC}$, and $\overset{\frown}{CA}$ have lengths 3,4, and 5, respectively. Find the area of triangle ABC.

Problem 4.51 (2002 AMC 12A) Triangle ABC is a right triangle with $\angle ACB$ as its right angle, $m\angle ABC = 60°$, and $AB = 10$. Let P be randomly chosen inside $\triangle ABC$, and extend $\overline{BP}$ to meet $\overline{AC}$ at D. What is the probability that $BD > 5\sqrt{2}$?

Problem 4.52 Let M be the intersection of line segments $\overline{AB}$ and $\overline{PQ}$. Show that $\dfrac{[PAB]}{[QAB]} = \dfrac{PM}{QM}$.

Problem 4.53 In the parallelogram $ABCD$, E is a point on $\overline{AD}$, F is a point on $\overline{AB}$, and $BE = DF$. Also $\overline{BE}$, $\overline{DF}$ intersect at G. Show that $\angle BGC = \angle DGC$.

Problem 4.54 ABC is a triangle with integer side lengths. Extend $\overline{AC}$ beyond C to point D such that $CD = 120$. Similarly, extend $\overline{CB}$ beyond B to point E such that $BE = 112$ and $\overline{BA}$ beyond A to point F such that $AF = 104$. If triangles CBD, BAE, and ACF all have the same area, what is the minimum possible area of triangle ABC?

Problem 4.55 Given an equilateral triangle ABC and a point P in the interior of $\triangle ABC$. Show that the sum of the distances from P to all three sides is equal to the altitude of $\triangle ABC$.

Problem 4.56 Let $ABCD$ by a square with side length 1. Let E, P, F be the midpoints of AD, CE, BP respectively. Find $[BFD]$.

Problem 4.57 Let P be an interior point in parallelogram $ABCD$, and $[APB] : [ABCD] = 2 : 5$. Find $[CPD] : [ABCD]$

Problem 4.58 Let $\odot O$ be a circle with radius 1. Let A be a point outside $\odot O$, and $OA = 2$. Let AB be tangent to $\odot O$ at B, and chord $BC \parallel OA$. Connect AC. Find the sum of the areas of $\triangle ABC$ and the region enclosed by BC and the minor arc $\overset{\frown}{BC}$.

Problem 4.59 Let E be the midpoint of BC in parallelogram $ABCD$, G be the intersection of AE and BD. If $[BEG] = 1$, find $[ABCD]$.

Problem 4.60 Let O be the intersection of the diagonals of convex quadrilateral $ABCD$. Given that $[ABC] = 5, [ACD] = 10$, and $[ABD] = 6$, find $[ABO]$.

Problem 4.61 (2002 AMC 12A) In triangle ABC, side $\overline{AC}$ and the perpendicular bisector of $\overline{BC}$ meet in point D, and $\overline{BD}$ bisects $\angle ABC$. If $AD = 9$ and $DC = 7$, what is the area of triangle ABD?

Problem 4.62 Let $\triangle ABC$ be a right triangle, where $\angle C$ is the right angle. Construct squares $ACDE$ and $BCFG$ on the outside of $\triangle ABC$. Assume that $\overline{AG}$ intersects $\overline{BC}$ at point P, and $\overline{BE}$ intersects $\overline{AC}$ at point Q, show that $CP = CQ$.

4.3 Circles

4.3.1 Fundamentals

Basic Definitions

- A *circle* is a collection of points of equal distance (called the *radius*) from a set point (called the *center*).
- Given two points A, B on a circle, the segment $\overline{AB}$ is called a *chord*.
- If a chord AB contains the center of the circle, we say A and B are *diametrically opposite*, and call AB a *diameter*.
- The portion of a circle that lies above or below a chord AB is called an *arc*. If the arc is more than half a circle it is called a *major arc*, less than half a circle is called a *minor arc*, and half a circle is called a *semicircle*. The arc will be denoted $\overarc{AB}$.
- Suppose $\overarc{AB}$ is an arc on a circle with center O. The *angular size* of the arc $\overarc{AB}$ is equal to the angle $\angle AOB$ (which is referred to as a *central angle*).
- Given a central angle $\angle AOB$ from an arc $\overarc{AB}$, the figure contained between the arc $\overarc{AB}$ and the radii $\overline{OA}, \overline{OB}$ is called a *sector*.

Measurements in Circles

- The area of a circle is given by πr^2 where r is the radius.
- The circumference of a circle is given by $2\pi r = \pi d$ where r, d are the radius and length of a diameter respectively.
- The *arc length* of $\overarc{AB}$ (that is, the distance walking from A to B along the circle) is given by $\dfrac{\theta}{360°} 2\pi r$ where θ is the angular size of $\overarc{AB}$ (measured in degrees).
- Similarly, the area of a sector from arc $\overarc{AB}$ is given by $\dfrac{\theta}{360°} \pi r^2$ where θ is the angular size of $\overarc{AB}$ (measured in degrees).

Theorem 4.4 (Chords and Bisectors)

In a circle:
- a radius is perpendicular to a chord if and only if the radius bisects the chord.
- the perpendicular bisector of a chord passes through the center of the chord.

Problem 4.63 Prove that in a circle, a radius is perpendicular to a chord if and only if the radius bisects the chord.

Problem 4.64   (a) Suppose two tires, each with radius 1ft rest upright on the ground and touching each other, as pictured below:

How much space is needed horizontally to store the tires?

(b) Repeat part (a) with two tires of radii 1 and 2 feet respectively.

Problem 4.65 Suppose you start with a circle of radius 1. Choose a point on the circle and draw another circle of radius 1 with the chosen point as its center. Let R denote the region consisting of all points that are inside both circles.

(a) Find the perimeter of R.

(b) Find the area of R.

Problem 4.66   (a) What is the radius of the largest circle that can fit in a quarter circle
          of radius 1?

    (b) What if instead you are fitting it into a 60°-sector?

Problem 4.67 Arrange 4 congruent circles so that (i) the center of the four circles form
a square with side length 10, and (ii) adjacent circles are tangent.

    (a) Find the radius of each circle.

    (b) Find the area of the region inside the square that is outside each of the circles.

Problem 4.68 Prove that in a circle, the perpendicular bisector of a chord passes through
the center of the circle.

Problem 4.69 Suppose you have three tires with radii $1, 2, 3$ feet respectively. You store them horizontally as in Problem 4.64 above. What is the minimum amount of horizontal space needed to store the tires? Justify your answer!

Problem 4.70 Suppose you start with a circle of radius 1. Choose a point on the circle and draw another circle of radius 1 with the chosen point as its center. Now pick one of the two intersection points and draw a third circle centered at this point, also with radius 1. Call R the region consisting of all points that are contained inside at least two circles.

(a) Find the perimeter of R.

(b) Find the area of R.

Problem 4.71 (a) What is the radius of the largest circle that can fit in a $120°$ sector of a circle of radius 1?

(b) What if you have a $240°$ sector?

4.3.2 Arcs and Angles

If points A, B, P are on a circle, we call $\angle APB$ an *inscribed angle*. The measure of $\angle APB$ is half the angular size of arc $\widehat{AB}$ (where the arc does NOT contain P).

Theorem 4.5

Suppose two chords (or their extensions) AC, BD intersect at a point P.
- If they intersect inside the circle as in the diagram below,

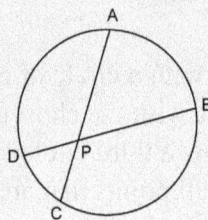

then $\angle APB$ is half the sum of the angular sizes of arcs $\widehat{AB}$ and $\widehat{CD}$.
- If they intersect outside the circle as in the diagram below,

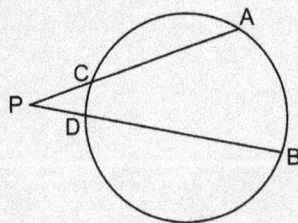

Then $\angle APB$ is half the difference of the angular sizes of arcs $\widehat{AB}$ and $\widehat{CD}$.

Problem 4.72 Let $\angle APB$ be an inscribed angle on a circle with center O. Prove that $\angle APB$ is half the angular size of arc $\widehat{AB}$ if:

(a) O lies on $\angle APB$.

(b) O lies inside $\angle APB$.

Problem 4.73 Prove that if two chords AC, BD intersect inside a circle at point P then the measure of $\angle APB$ is half the sum of the angular sizes of $\overset{\frown}{AB}, \overset{\frown}{CD}$.

Problem 4.74 Suppose $\overset{\frown}{AB}$ is an arc with angular size $60°$ and CD is a diameter such that if rays $\overrightarrow{BA}, \overrightarrow{DC}$ are extended to intersect at a point E, $\angle AEC = 30$. Find the angular size of arc $\overset{\frown}{BD}$.

Problem 4.75 Suppose two perpendicular chords intersect and divide each other in a ratio of $1 : 2$. Find the radius of the circle if each chord is 12in long.

Problem 4.76 Suppose ω is a circle with radius 6 and center O. Let $\overset{\frown}{AB} = 135°$. Let C be on ω such that $\overline{OA} \parallel \overline{BC}$. Find $[OACB]$.

Problem 4.77 Suppose you have a regular hexagon with side length 6. Let G, H, I be midpoints of every other side of the hexagon. Let $\widehat{GH}, \widehat{HI}, \widehat{IG}$ be arcs of three non-overlapping circles with radii 9. Find the area of the region in the hexagon in between all three arcs.

Problem 4.78 Let $\angle APB$ be an inscribed angle on a circle with center O. Prove that $\angle APB$ is half the angular measure of arc $\widehat{AB}$ if O lies outside $\angle APB$.

Problem 4.79 Prove that if two chords AC, BD intersect outside a circle at point P then the measure of $\angle APB$ is half the difference of the angular sizes of $\widehat{AB}, \widehat{CD}$.

Problem 4.80 Suppose A, B, C are points on a circle such that the angular measures of arc $\widehat{AB}$ (not containing C) and arc $\widehat{CA}$ (not containing B) are in ratio $5:9$. Suppose further that $\angle ABC = 90°$. Find the measure of $\angle BAC$.

Problem 4.81 Suppose $\overline{AB}, \overline{CD}$ are two chords of equal length who intersect at E. Suppose $\angle AED = 120°$, and $AE : EB = CE : ED = 1 : 2$. Further, suppose $AC = 2$.

   (a) Find the distance from E to the center of the circle.

   (b) Find the radius of the circle.

Problem 4.82 Recall the construction of trapezoid $OACB$ as in Problem 4.76 What angle does $\overset{\frown}{AB}$ have to be so that $ABCD$ is a parallelogram?

4.3.3 Power of a Point

Theorem 4.6 Power of a Point

- Let $\overline{AB}, \overline{CD}$ be chords, which intersect at E inside the circle. Then $AE \cdot BE = CE \cdot DE$.
- Let $\overline{AB}, \overline{CD}$ be chords, which are extended to intersect at E outside the circle. Then $AE \cdot BE = CE \cdot DE$.
- Let a line tangent to the circle at C intersect the extension of chord $\overline{AB}$ at E. Then $AE \cdot BE = CE^2$. (Note: Think of this as the previous case with $C = D$.)

Problem 4.83   (a) Suppose two inscribed angles share the same arc. Prove that the angles are equal.

(b) Prove the Power of a Point formula for chords intersecting inside the circle.

Problem 4.84 Suppose a line is tangent to the circle at C and it intersects a diameter $\overline{AB}$ at E. Prove that $CE^2 = AE \cdot BE$.

Problem 4.85 Suppose chords $\overline{AB}, \overline{CD}$ intersect at E, such that $AE : EB = 1 : 3$ and $CE : ED = 1 : 12$. Find the ratio of $AB : CD$.

Problem 4.86 Suppose we have a rectangle $ABCD$ with $AB = 8$, $BC = 12$. "Inscribe" a circle in the rectangle so that it touches sides $\overline{AB}, \overline{BC}, \overline{AD}$. Let M be the midpoint of $\overline{AB}$. Call $E \neq M$ the intersection of $\overline{MD}$ with the circle. Find DE.

Problem 4.87 Prove the Power of a Point formula for two chords whose extensions intersect outside a circle.

Problem 4.88 Prove the Power of a Point formula for arbitrary chord $\overline{AB}$ which intersects a tangent line (tangent to the circle at C) at E.

Problem 4.89 Suppose diameter $\overline{CD}$ intersects chord $\overline{AB}$ at E, so that $AE = 4, EB = 9$. If the diameter of the circle is 15, Find CE and ED.

Problem 4.90 Recall the setup from Problem 4.86 Assume that CE is tangent to the circle at E (It is, as a challenge try to prove it!). Find the length of CE.

4.4 **Solid Geometry**

Points in three dimensions can be thought of as using x, y, z coordinates, written (x, y, z). We have analogues of the Pythagorean Theorem and Distance Formula in three dimensions: The distance from $(0,0,0)$ to (a,b,c) is $\sqrt{a^2 + b^2 + c^2}$.

Given two distinct lines, there are three possibilities: (i) they are parallel, (ii) they intersect, or (iii) they are *skew*. Given two distinct planes, there are two possibilities: (i) they are parallel, or (ii) they intersect (and their intersection is a line).

It is useful to be comfortable with the volume and surface area of various solids. The following information is often useful.

- **Sphere**: The collection of points (in three dimensions) of equal distance (called the *radius* from a *center*).
  - A sphere with radius r has volume $\dfrac{4}{3}\pi r^3$.
  - A sphere with radius r has surface area $4\pi r^2$.
- **Cube**: The 3-D version of a square.
  - A cube with side length s has volume s^3.
  - A cube with side length s has surface area $6s^2$.
  - A cube has 8 vertices, 12 edges, and 6 faces. The 6 faces are all squares.
- **Rectangular Prism**: A "box".
  - A rectangular prism with sides l, w, h (often 'length', 'width', 'height') has volume lwh.
  - A rectangular prism with sides l, w, h has surface area $2(lw + wh + lh)$.
  - Like a cube, a rectangular prism has 8 vertices, 12 edges, and 6 faces. The 6 faces are all rectangles.
- **Cylinder**: A "can".
  - A cylinder with height h and radius r has volume $\pi r^2 h$.
  - A cylinder with height h and radius r has surface area $2\pi r^2 + 2\pi rh$.
- **(General) Right Prism**: Similar to rectangular prism or cylinders with an arbitrary polygon as the base (the 'top' and 'bottom').
  - A right prism with height h and whose base has area B has volume Bh.
  - A right prism with height h and base with area B and perimeter P has surface area $2B + Ph$.
  - If the base polygon has n sides, the right prism has $2n$ vertices, $3n$ edges, and $n + 2$ faces. Every side face is rectangular and perpendicular to the bases.
  - **Note**: The volume formula holds even if the prism is not right, that is if the two bases are parallel but not necessarily 'above' each other.
- **Square Right Pyramid**: The standard "pyramid from Egypt" solid, with a square *base* and *apex* (or top point) that is centered above the square.
  - A square right pyramid with height h with square base of side length s has

volume $\frac{1}{3}s^2h$.

- ◦ A square right pyramid with height h with square base of side length s has surface area $s^2 + 2sL$, where $L = \sqrt{h^2 + s^2/4}$ (L is called the *slant height*).
- ◦ A square right pyramid has 5 vertices, 8 edges, and 5 faces. The 4 side faces are all triangles.

- **Right Cone**: The standard "ice cream cone" solid, with a circular *base* and *apex* that is centered above the circle.

  - ◦ A right with height h with square base of radius r has volume $\frac{1}{3}\pi r^2 h$.
  - ◦ A square right pyramid with height h with square base of side length s has surface area $\pi r^2 + \pi rL$, where $L = \sqrt{h^2 + r^2}$ (L is called the *lateral height*).

- **(General) Right Pyramid**: Similar to a square right pyramid or right cone with an arbitrary polygon as the base.

  - ◦ A right pyramid with height h and whose base has area B has volume $\frac{1}{3}Bh$.
  - ◦ A right pyramid with height h and base with area B and perimeter P has surface area $B + PL/2$, where $L = \sqrt{h^2 + r^2}$, where r is the inradius of the base (L is called the *slant height*).
  - ◦ If the base polygon has n sides, the right pyramid has $n+1$ vertices, $2n$ edges, and $n+1$ faces. The n side faces are all triangles.
  - ◦ **Note**: The volume formula holds even if the pyramid is not right, that is if the apex is not necessarily centered above the base.

- **Tetrahedron**: A triangular pyramid. Alternatively a solid made up of four triangles.

  - ◦ As a tetrahedron is a triangular pyramid, the above formulas hold.
  - ◦ In particular, a *regular* tetrahedron is a tetrahedron made up of 4 equilateral triangles.

Problem 4.91   (a) Find the volume of the largest sphere that fits in a cube of volume 1. (That is, inscribe a sphere inside the cube.)

(b) Find the volume of the smallest sphere that holds a cube of volume 1. (That is, circumscribe a sphere outside the cube.)

Problem 4.92 (2010 AMC 10A) Suppose we have a cube with side length 4. In the middle of each face of the cube, cut a 2 by 2 square hole all the way through the cube. What is the volume of the remaining solid after all the holes are cut.

Problem 4.93 Suppose you have a regular tetrahedron with side length a.
(a) Show that the surface area is $a^2\sqrt{3}$.

(b) Show that the volume is $\dfrac{a^3\sqrt{2}}{12}$.

Problem 4.94 Suppose $S - ABC$ is a regular tetrahedron with apex S. Cut off the top "half" of the tetrahedron (that is, cut through the midpoints of $\overline{SA}, \overline{SB}, \overline{SC}$ and leave the bottom).
(a) How many vertices, edges, and faces does the resulting solid have?

(b) Find the volume and the surface area as ratios to the original volume and surface area. Hint: Try to do so without using Problem 4.93

Problem 4.95 (2014 AMC 10A) Stack cubes with side length $1, 2, 3, 4$ as shown in the diagram below.

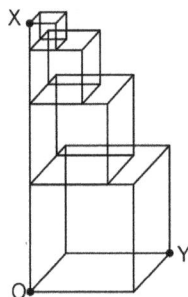

(a) Find the distance from X to Y.

(b) Find the length of the portion of $\overline{XY}$ contained in the bottom square.

Problem 4.96 (2010 AMC 10A) Suppose a bored bee lives on a cube with side length 1. For "fun" he decides to visit every vertex of the cube, each exactly once, starting and ending at the same vertex. It will travel from one vertex to another using straight lines (either crawling or flying). Give an example of a path that uses the maximum distance and find this distance.

Problem 4.97   (a) Four identical balls (spheres), each of radius 1in, are glued to the ground so that their centers form the vertices of a square with side length 2in. Suppose you rest a fifth identical ball on the four balls (so the fifth ball is a sphere externally tangent to the other spheres). How far does this ball rest off the ground?

(b) Repeat the setup of (a). Suppose now, however, that the fifth ball is not the same size as the others, and that it rests 1in off the ground. Find the radius of the fifth ball.

Problem 4.98   (a) Find the volume of the largest sphere that can fit inside a cone of radius 1 and height $\sqrt{3}$.

(b) Assume the sphere as in (a) is placed inside the cone. Suppose you now want to fit another sphere in the cone that is tangent to the base. Find the radius of the largest such sphere.

Problem 4.99 Find the surface area of the remaining solid in Problem 4.92

Problem 4.100 Suppose you pick 4 vertices of a cube to form a tetrahedron.

   (a) How many different (non-congruent) tetrahedrons are possible? Are any of them regular tetrahedrons?

   (b) Find the volumes for each of the possibilities in (a) if the cube has volume 1.

Problem 4.101 Suppose you have a regular square pyramid $S - ABCD$ with height 6 whose square has side length 4. Call the midpoints of the square E, F, G, H (on $\overline{AB}, \overline{BC}, \overline{CD}, \overline{DA}$ respectively) and the midpoints of $\overline{SA}, \overline{SB}, \overline{SC}, \overline{SD}$ respectively T, U, V, W. Form polyhedron $EFGH - TUVW$ (with 10 faces).

   (a) Describe the faces of $EFGH - TUVW$. How many vertices and edges does the polyhedron have?

   (b) Find the volume of $EFGH - TUVW$. Hint: Do this indirectly.

Problem 4.102 Suppose solid cubes of side length 1 are removed from every corner of a solid cube with side length 3.

(a) What is the volume and surface area of the new solid?

(b) How many vertices, edges, and faces does the new solid have?

Problem 4.103 Suppose you have a unit cube. Pick two opposite corners. In each corner, form a tetrahedron using the corner and the three adjacent vertices. Remove these two tetrahedrons and call the resulting polyhedron $\mathscr{S}$.

(a) How many vertices, edges, and faces does the resulting polyhedron have? Describe the faces.

(b) Find the volume of $\mathscr{S}$.

(c) Suppose $\mathscr{S}$ is resting on one of the faces (ignore whether the polyhedron would actually balance or not). What are different possible "heights" of $\mathscr{S}$?

Problem 4.104 Stack cubes with side length $1, 2, 3, 4$ as shown in the diagram below.

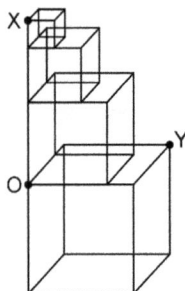

(a) Find the distance from X to Y.

(b) Find the length of the portion of $\overline{XY}$ *not* contained inside the cubes.

Problem 4.105   (a) Repeat Problem 4.96 if the bee wants the shortest path.

(b) Find the maximum and minimum paths if the bee instead lives on a Right Prism with height 2 whose base is a regular hexagon with side length 1.

Problem 4.106 (2013 AMC 10A) Suppose 6 spheres of radius 1 are arranged so that the centers form a regular hexagon with side length 2. All 6 spheres are internally tangent to a larger 7th sphere whose center is the center of the hexagon. Lastly, an 8th sphere is externally tangent to the 6 smaller spheres and internally tangent to the large sphere.

(a) Find the radius of the large sphere.

(b) Find the radius of the 8th sphere.

Problem 4.107 Suppose you have tetrahedron $S - ABC$. Cut the tetrahedron with a plane between S and $\triangle ABC$, forming a new tetrahedron $S - A'B'C'$ (with A' on $\overline{SA}$, etc.). Prove that the ratio of volumes of $S - ABC$ to $S - A'B'C'$ is equal to $SA \cdot SB \cdot SC :$ $SA' \cdot SB' \cdot SC'$.

Problem 4.108 (2015 AMC 10B #17) Suppose you have a right rectangular prism with length, width, height equal to $3, 4, 5$. Connect all the centers of the faces to form an octahedron (a polyhedron with 8 sides). What is the octahedron's volume?

Problem 4.109 Suppose you start with a right cone and cut off the top of the cone with a plane parallel to the base. The resulting solid, called a *frustum* has two circular "bases", say with radii R and r (with $R > r$), and height h. (Hence from the side the frustum looks like a trapezoid with bases R, r and height H.)

(a) Show that the volume of a frustum is $\dfrac{\pi H}{3}(R^2 + Rr + r^2)$.

(b) Suppose a sphere can be inscribed in a frustum with base radii r, R such that the sphere is tangent to the two bases and the side. Find the radius of such a sphere in terms of r, R.

Problem 4.110 Start with right triangular prism $ABC - DEF$. Divide the prism into four parts using the planes through points A, B, F and D, E, C. Find the ratio of the volumes of these four parts.

Problem 4.111 Suppose you have a right pyramid with a regular hexagon as a base. Suppose further that the surface area of the pyramid is 3 times the area of the base. If the volume of the pyramid is $6\sqrt{3}$, find side length of the hexagon.

4.5 Trigonometry

The main trigonometric functions (and their inverses) are listed below.

Function	Domain	Range
$\sin x$	$(-\infty, +\infty)$	$[-1, 1]$
$\cos x$	$(-\infty, +\infty)$	$[-1, 1]$
$\tan x$	$x \neq \dfrac{\pi}{2} + k\pi$	$(-\infty, +\infty)$
$\cot x$	$x \neq k\pi$	$(-\infty, +\infty)$
$\sec x$	$x \neq \dfrac{\pi}{2} + k\pi$	$(-\infty, -1] \cup [1, +\infty)$
$\csc x$	$x \neq k\pi$	$(-\infty, -1] \cup [1, +\infty)$
$\arcsin x$	$[-1, 1]$	$\left[-\dfrac{\pi}{2}, \dfrac{\pi}{2}\right]$
$\arccos x$	$[-1, 1]$	$[0, \pi]$
$\arctan x$	$(-\infty, +\infty)$	$\left(-\dfrac{\pi}{2}, \dfrac{\pi}{2}\right)$
$\text{arccot}\, x$	$(-\infty, +\infty)$	$(0, \pi)$

Values of these functions can be calculated using the mnemonic SOH-CAH-TOA (sine is opposite over hypotenuse, cosine is adjacent over hypotenuse, tangent is opposite over adjacent) and the special angles. Common values are summarized below.

Rad	0	$\pi/6$	$\pi/4$	$\pi/3$	$\pi/2$	$2\pi/3$	$3\pi/4$	$5\pi/6$	π
Deg	0	30	45	60	90	120	135	150	180
$\sin x$	0	$1/2$	$\sqrt{2}/2$	$\sqrt{3}/2$	1	$\sqrt{3}/2$	$\sqrt{2}/2$	$1/2$	0
$\cos x$	1	$\sqrt{3}/2$	$\sqrt{2}/2$	$1/2$	0	$-1/2$	$-\sqrt{2}/2$	$-\sqrt{3}/2$	-1
$\tan x$	0	$\sqrt{3}/3$	1	$\sqrt{3}$	-	$-\sqrt{3}$	-1	$-\sqrt{3}/3$	0
$\cot x$	-	$\sqrt{3}$	1	$\sqrt{3}/3$	0	$-\sqrt{3}/3$	-1	$-\sqrt{3}$	-
$\sec x$	1	$2\sqrt{3}/3$	$\sqrt{2}$	2	-	-2	$-\sqrt{2}$	$-2\sqrt{3}/3$	-1
$\csc x$	-	2	$\sqrt{2}$	$2\sqrt{3}/3$	1	$2\sqrt{3}/3$	$\sqrt{2}$	2	-

The "Magic Hexagon (Wheel) of Trigonometric Identities" is very useful for helping to remember various identities involving trigonometric functions.

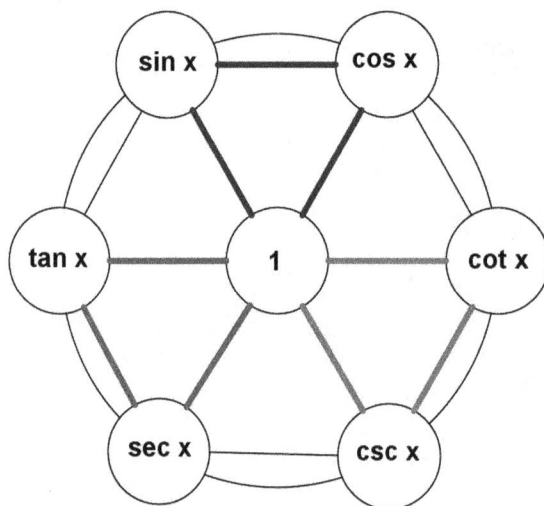

1. **On the diagonals—the reciprocal identities**

$$\sin x = \frac{1}{\csc x}, \qquad \cos x = \frac{1}{\sec x}, \qquad \tan x = \frac{1}{\cot x}$$

2. **On the upside-down triangles—the pythagorean identities**

$$\begin{aligned} \sin^2 x + \cos^2 x &= 1 \\ \tan^2 x + 1 &= \sec^2 x \\ 1 + \cot^2 x &= \csc^2 x \end{aligned}$$

3. **On the circle—the quotient identities** Any vertex is equal to the quotient of the next two consecutive vertices. It works in both directions.

$$\tan x = \frac{\sin x}{\cos x}, \qquad \sin x = \frac{\cos x}{\cot x}, \qquad \sec x = \frac{\tan x}{\sin x}, \qquad \text{etc.}$$

There are many other identities involving trigonometric functions. Some of the most important are summarized below.

1. **Co-functions (complementary angles)**

$$\sin\left(\frac{\pi}{2} - x\right) = \cos x, \qquad \cos\left(\frac{\pi}{2} - x\right) = \sin x, \qquad \tan\left(\frac{\pi}{2} - x\right) = \cot x$$

2. **Even-Odd properties**

$$\sin(-x) = -\sin x, \qquad \cos(-x) = \cos x, \qquad \tan(-x) = -\tan x$$

3. **Periodic properties**

$$\sin(x+2\pi) = \sin x, \qquad \cos(x+2\pi) = \cos x, \qquad \tan(x+\pi) = \tan x$$

4. **Supplementary angles**

$$\sin(\pi - x) = \sin x, \qquad \cos(\pi - x) = -\cos x, \qquad \tan(\pi - x) = -\tan x$$

5. **Other properties**

$$\sin(x+\pi) = -\sin x, \qquad \cos(x+\pi) = -\cos x$$

$$\sin(2\pi - x) = -\sin x, \qquad \cos(2\pi - x) = \cos x, \qquad \tan(2\pi - x) = -\tan x$$

6. **Sum & difference formulas**

$$\begin{aligned}
\sin(x \pm y) &= \sin x \cos y \pm \cos x \sin y \\
\cos(x \pm y) &= \cos x \cos y \mp \sin x \sin y \\
\tan(x \pm y) &= \frac{\tan x \pm \tan y}{1 \mp \tan x \tan y}
\end{aligned}$$

7. **Double angle formulas**

$$\begin{aligned}
\sin 2x &= 2 \sin x \cos x \\
\cos 2x &= \cos^2 x - \sin^2 x \\
&= 2 \cos^2 x - 1 \\
&= 1 - 2 \sin^2 x \\
\tan 2x &= \frac{2 \tan x}{1 - \tan^2 x}
\end{aligned}$$

8. **Power-reducing/Half angle formulas**

$$\begin{aligned}
\sin^2 x &= \frac{1 - \cos 2x}{2}, & \cos^2 x &= \frac{1 + \cos 2x}{2} \\
\tan^2 x &= \frac{1 - \cos 2x}{1 + \cos 2x}, & \tan x &= \frac{\sin 2x}{1 + \cos 2x} = \frac{1 - \cos 2x}{\sin 2x}.
\end{aligned}$$

9. **Product-to-sum formulas**

$$\begin{aligned}
\sin x \cos y &= \frac{1}{2}\left(\sin(x+y) + \sin(x-y)\right) \\
\cos x \sin y &= \frac{1}{2}\left(\sin(x+y) - \sin(x-y)\right) \\
\cos x \cos y &= \frac{1}{2}\left(\cos(x+y) + \cos(x-y)\right) \\
\sin x \sin y &= -\frac{1}{2}\left(\cos(x+y) - \cos(x-y)\right)
\end{aligned}$$

10. **Sum-to-product formulas**

$$\sin x + \sin y = 2\sin\left(\frac{x+y}{2}\right)\cos\left(\frac{x-y}{2}\right)$$

$$\sin x - \sin y = 2\cos\left(\frac{x+y}{2}\right)\sin\left(\frac{x-y}{2}\right)$$

$$\cos x + \cos y = 2\cos\left(\frac{x+y}{2}\right)\cos\left(\frac{x-y}{2}\right)$$

$$\cos x - \cos y = -2\sin\left(\frac{x+y}{2}\right)\sin\left(\frac{x-y}{2}\right)$$

11. **(Extended) Law of Sines**
    For a triangle ABC with circumradius R,

$$\frac{a}{\sin A} = \frac{b}{\sin B} = \frac{c}{\sin C} = 2R$$

12. **Law of Cosines**
    For a triangle ABC,

$$
\begin{aligned}
a^2 &= b^2 + c^2 - 2bc\cos A \\
b^2 &= c^2 + a^2 - 2ca\cos B \\
c^2 &= a^2 + b^2 - 2ab\cos C
\end{aligned}
$$

Problem 4.112 Compute the following values.

(a) $\cos 225°$

(b) $\sin\dfrac{11\pi}{6}$

(c) $\tan 405°$

(d) $\cot\dfrac{9\pi}{8}$

(e) $\sin 75°$

Problem 4.113 Find a formula for:

(a) $\sin(x+90°)$

(b) $\cos(x+90°)$

Problem 4.114 Evaluate the following.

(a) $\cos\left(\dfrac{\pi}{3}\right)\cos\left(-\dfrac{\pi}{12}\right) - \sin\left(\dfrac{\pi}{3}\right)\sin\left(-\dfrac{\pi}{12}\right)$

(b) $\sin\left(2\arccos\dfrac{4}{5}\right)$

(c) $\cos\left(\arccos\left(-\dfrac{\sqrt{2}}{2}\right)-\dfrac{\pi}{2}\right)$

(d) $\cos^4\dfrac{\pi}{24}-\sin^4\dfrac{\pi}{24}$

(e) $\cot 70^\circ + 4\cos 70^\circ$

(f) $\cos\dfrac{\pi}{15}\cos\dfrac{2\pi}{15}\cos\dfrac{4\pi}{15}\cos\dfrac{8\pi}{15}$

(g) $(1-\cot 23^\circ)(1-\cot 22^\circ)$

(h) $\arctan \dfrac{1}{2} + \arctan \dfrac{1}{3}$

(i) $\cos^2 x + \cos^2 \left(x + \dfrac{2\pi}{3} \right) + \cos^2 \left(x + \dfrac{4\pi}{3} \right)$

Problem 4.115 Simplify:

$$\cos(A+B)\cos B + \sin(A+B)\sin B$$

Problem 4.116 Simplify:

$$\frac{\cos(180° + t)\sin(t + 360°)}{\sin(-t - 180°)\cos(-180° - t)}$$

Problem 4.117 Simplify:

$$\cfrac{1}{1-\cfrac{1}{1-\cfrac{1}{1-\sin^2 x}}}$$

Problem 4.118 Given that $\csc x + \cot x = 5$, find the value of $\csc x - \cot x$.

Problem 4.119 Given that $\tan y = -\sqrt{5}$, find the value of $\dfrac{\sqrt{5}\sin y + 2\cos y}{\cos y - \sqrt{5}\sin y}$.

Problem 4.120 Given that $\tan y = -\sqrt{5}$, find the value of $(\sin y - \cos y)^2$.

Problem 4.121 The quadratic equation $2x^2 - (\sqrt{3}+1)x + m = 0$ has two roots $\sin\theta$

and $\cos\theta$, find the value of $\dfrac{\sin\theta}{1-\cot\theta}+\dfrac{\cos\theta}{1-\tan\theta}$.

Problem 4.122 Given that $\sin\alpha-\cos\alpha=-\dfrac{\sqrt{5}}{5}$, and $180°<\alpha<270°$, find the value of $\tan\alpha$.

Problem 4.123 Given that $\sin x=\dfrac{4}{5}$, where $x\in\left(\dfrac{\pi}{2},\pi\right)$, and $\cos y=-\dfrac{5}{13}$, where $y\in\left(\pi,\dfrac{3\pi}{2}\right)$. Find the value of $\cos(x+y)$.

Problem 4.124 Find maximum and minimum values of

(a) $\dfrac{1}{2}\cos x-\dfrac{\sqrt{2}}{2}\sin x$.

(b) $3\sin x+4\cos x$

(c) $a\sin x + b\cos x$, where $ab \neq 0$.

(d) $\sin x + \sin\left(x + \dfrac{\pi}{4}\right)$

Problem 4.125 Given $x, y \in \left[-\dfrac{\pi}{4}, \dfrac{\pi}{4}\right]$, and $a \in \mathbb{R}$, and

$$\begin{cases} x^3 + \sin x - 2a &= 0 \\ 4y^3 + \dfrac{1}{2}\sin 2y + a &= 0 \end{cases}$$

Find $\cos(x + 2y)$.

Problem 4.126 Let $0 < x < \dfrac{\pi}{4}$, arrange the following four numbers in increasing order:

$$(\tan x)^{\tan x}, \quad (\tan x)^{\cot x}, \quad (\cot x)^{\tan x}, \quad (\cot x)^{\cot x}$$

Problem 4.127 If $\sin x = -\dfrac{4}{5}$ and $\tan x < 0$, find the exact value of $\cos 3x$.

Problem 4.128 Given obtuse angles α and β, satisfying: $\sin \alpha = \dfrac{12}{13}$, $\cos(\beta - \alpha) = \dfrac{3}{5}$. Find $\sin \beta$.

Problem 4.129 Find the smallest positive period of $f(x) = |\sin x| + |\cos x|$.

Problem 4.130 Simplify:

(a) $3(\sin^4 x + \cos^4 x) - 2(\sin^6 x + \cos^6 x)$

(b) $\sqrt{\sin^4 x + 4\cos^2 x} - \sqrt{\cos^4 x + 4\sin^2 x}$

(c) $\cos x + \cos 2x + \cos 3x + \cdots + \cos nx$

(d) $\displaystyle\sum_{k=0}^{n} \arctan \frac{1}{k^2 + k + 1}$

(e) $\displaystyle\sum_{k=1}^{n} \arctan \frac{1}{2k^2}$

Problem 4.131 Let x and y be real numbers such that

$$\sin x + \sin y = \frac{\sqrt{2}}{2} \quad \text{and} \quad \cos x + \cos y = \frac{\sqrt{6}}{2},$$

find the value of $\sin(x+y)$.

Problem 4.132 Solve the following inequality for $0 \le x \le 2\pi$:

$$1 + \tan x > 2\cos x + 2\sin x$$

Problem 4.133 Solve the following equations:

(a) $\sqrt{3}\sin x = \cos x$

(b) $3\cos^2 x = \sin x \sin 2x$

(c) $\tan x - 3\cot x = 0$, for $x \in [0, 2\pi]$.

(d) $\sin^2 x + 2\sin x \cos x = 3\cos^2 x$

5. Combinatorics

5.1 Combinatorics Fundamentals

5.1.1 Counting Basics

Sequential Counting Principle (Product Rule)

Suppose that a procedure can be broken down into k successive tasks. If there are n_1 ways to do the first task, and n_2 ways to do the second task after the first task has been done, and so on, then there are $n_1 \times n_2 \times \cdots \times n_k$ ways to do the procedure.

Additive Counting Principle (Sum Rule)

Suppose we have tasks $T_1, T_2, \ldots, T_k$ that can be done in $n_1, n_2, \ldots, n_k$ ways, respectively, and no two of these tasks can be done at the same time, then there are $n_1 + n_2 + \ldots + n_k$ ways to do one of these tasks.

> **Remark**
>
> Both of these rules are "reversible". For example, suppose a procedure can be broken down into 2 successive tasks. If there are n ways to do the entire procedure and m ways to do the 2nd task, then there are $\dfrac{n}{m}$ ways of doing the 1st task.

Example 5.1

Suppose John has 2 hats, 5 shirts, 2 jackets, 4 pairs of pants, 5 pairs of shorts, and 3 pairs of shoes.

(a) Suppose John makes an outfit consisting of a shirt, a pair of pants, and a pair of shoes. How many different outfits does he have?
   Solution: 60.
   We use the product rule: $5 \cdot 4 \cdot 3 = 60$.

(b) Repeat (a) if John *can* wear shorts instead of pants.
   Solution: 135.
   Using the Sum Rule, we have $4 + 5 = 9$ choices for leg wear. We then proceed as in (a): $5 \cdot 9 \cdot 3 = 135$.

(c) Now suppose John can wear shorts or pants as in (b), *but* if he wears shorts, he will also wear a hat and possibly a jacket.
   Solution: 510.
   Consider the two cases (so Sum Rule) based on whether John wears shorts or pants. If he wears pants, it is the same as (a); if he wears shorts, we also have to choose which hat he wears and which (if any) jacket he wears ($2 + 1 = 3$ choices for jacket). The total is: $5 \cdot 4 \cdot 3 + 5 \cdot 5 \cdot 3 \cdot 2 \cdot 3 = 510$.

Permutations

Permutation means arrangement of things *in a certain order*. The number of permutations of r elements taken out of a set of n elements (without repeating) is denoted $_nP_r$:

$$_nP_r - n(n-1)(n-2)\cdots(n-r+1) - \frac{n!}{(n-r)!}.$$

Combinations

Combination means selection of things where *order does not matter*. The number of combinations of r elements taken out of a set of n elements is denoted $_nC_r$ or $\binom{n}{r}$:

$$_nC_r = \binom{n}{r} = \frac{n(n-1)(n-2)\cdots(n-r+1)}{r!} = \frac{n!}{r!(n-r)!}.$$

Example 5.2

Suppose you have 10 people at a party.
- How many ways are there to line everyone up for a picture?
  Solution: $10! = 3628800$.
  This is a permutation ($_{10}P_{10}$).
- How many ways can we line up 4 of the 12 people for a picture?
  Solution: $\dfrac{10!}{6!} = 5040$.
  This is a permutation ($_{10}P_4$).
- How many ways are there to pick a group of 6 people to play a game of charades together?
  Solution: $\dbinom{10}{6} = 210$.
  This is a combination.

Problem 5.1 How many factors of 2^{95} are larger than $1,000,000$?

Problem 5.2 Telephone numbers in *Land of Nosix* have 7 digits, and the only digits available are $\{0,1,2,3,4,5,7,8\}$. No telephone number may begin in $0,1$, or 5. Find the number of telephone numbers possible that meet the following criteria:

(a) you may have repeated digits.

(b) you may not have repeated digits.

(c) you may have repeated digits, but the phone number must be even.

(d) you may not have repeated digits, and the phone number must be odd.

Problem 5.3 Suppose you have a group of 12 people. How many different photographs are there of everyone lined up if:

(a) all the people look different?

(b) 4 of the people are identical quadruplets who have dressed identically?

(c) 3 of the people are a family and must stand next to each other?

(d) 3 of the people do not get along, and cannot all 3 be right next to each other in a group?

Problem 5.4 Suppose you have a student group with 15 males and 10 females.

(a) How many ways are there to pick a group of 5 males and 5 females?

(b) How many ways are there to pick an Executive Committee of 5 members and a Party Planning Committee of 5 members? Members can be on both committees at once, but each committee must have at least one male and at least one female.

(c) Suppose you still need to pick an Executive Committee and Party Planning Committee (each with 5 members). This time, only the Executive Committee is required to have a member of each gender, but now members are *not* allowed to be on both committees at once.

Problem 5.5 The number 3 can be expressed as a sum of one or more positive integers in four ways, namely, as $3, 1+2, 2+1$, and $1+1+1$.

(a) How many ways can the number 6 be expressed as the sum of one or more positive integers less than or equal to 2?

(b) How many total ways can 6 be expressed as the sum of one or more positive integers?

Problem 5.6 A bookshelf contains 5 German books, 7 Spanish books, and 8 French books. Each book is different from one another.

(a) How many different arrangements can be done of these books?

(b) How many different arrangements can be done if books of each language must be next to each other?

(c) How many different arrangements can be done if no two German books must be next to each other?

5.1.2 Advanced Counting

Principle of Inclusion-Exclusion

The *Principle of Inclusion-Exclusion* (or PIE for short) helps calculate the size of the union of two or more sets.

- For two sets, we have $n(A \cup B) = n(A) + n(B) - n(A \cap B)$. (Here $n(A)$ denotes the size of A.)
- For three sets, we have $n(A \cup B \cup C) = n(A) + n(B) + n(C) - n(A \cap B) - n(A \cap C) - n(B \cap C) + n(A \cap B \cap C)$.
- Venn Diagrams are useful in remembering and visualizing the Principle of Inclusion-Exclusion.

Example 5.3

Using the PIE formula, find the number of positive integers between 1 and 1000 that are either a multiple of 5, a multiple of 6, or a multiple of 7.

Solution: $200 + 166 + 142 - 33 - 28 - 23 + 4 = 428$.

Let A be the multiples of 5, B the multiples of 6, and C the multiples of 7. We then have $n(A) = 200, n(B) = 166, n(C) = 142$. Note that $A \cap B$ is all the multiples of 30, $A \cap C$ multiples of 35, $B \cap C$ multiples of 42, so $n(A \cap B) = 33, n(A \cap C) = 28, n(B \cap C) = 23$. Lastly, $A \cap B \cap C$ is the multiples of 210, so $n(A \cap B \cap C) = 4$. Hence, $n(A \cup B \cup C) = 200 + 166 + 142 - 33 - 28 - 23 + 4 = 428$.

Stars and Bars

Stars and Bars (Balls and Urns, etc., there are many different names) is a counting technique that should be memorized as all costs.

Theorem 5.1 Stars and Bars (Non-Negative Version)

Given a positive integer n, and positive integer k, the number of ways to express n as the sum of k non-negative integers ($n = a_1 + a_2 + \cdots + a_k$ where $a_1, a_2, \ldots, a_k$ are non-negative) is $\binom{n+k-1}{n} = \binom{n+k-1}{k-1}$.

Theorem 5.2 Stars and Bars (Positive Version)

Given a positive integer n, and positive integer k, the number of ways to express n as the sum of k positive integers ($n = a_1 + a_2 + \cdots + a_k$ where $a_1, a_2, \ldots, a_k$ are positive) is $\binom{n-1}{k-1}$.

Remark

Both versions above can be easily proved from one another. It is useful to know how to prove the positive version from the non-negative version.

Example 5.4

Beagel likes bagels, and he went to the Bagel Shop to buy 6 bagels for breakfast. The Bagel Shop sells 3 types of bagels: sourdough, blueberry, and sesame seeds.

(a) If Beagel plans to buy at least one of each type. In how many ways can he do this?
   Solution: 10.
   $a + b + c = 6$ and a, b, c are all positive integers. This is a Stars and Bars problem, and the answer is $\binom{5}{2} = 10$. The answer can also be obtained by listing all the possibilities.

(b) If Beagel does not need to buy at least one of each type, how many ways can he buy the 6 bagels?
   Solution: $\binom{6+3-1}{6} = 28$.
   Now we have $a + b + c = 6$ were a, b, c are all non-negative integers. It is still Stars and Bars, but the alternate version.

(c) How many ways are there to buy the bagels so that Beagel gets at least 2 types?
   Solution: $\binom{6+3-1}{6} - 3 = 25$.
   There are only 3 of the total outcomes where Beagel does not have at least 2 types.

Problem 5.7 Suppose you have 40 identical balls and 8 numbered boxes. How many ways are there to put the balls into the boxes if:

(a) there are no restrictions?

(b) each box has at least two balls?

(c) no box has more than 5 balls?

(d) the first box has exactly 10 balls?

Problem 5.8 Suppose you have 10 blue, 10 red, and 10 green balls. You want to arrange the balls so that no two green balls are next to each other. How many ways are there to do this if

(a) the balls are in a row and each ball is numbered?

(b) the balls are in a row and each ball is identical?

(c) the balls are in a circle and each ball is numbered?

Problem 5.9 Given positive integers $1, 2, 3, \ldots, n$. Let a permutation of these numbers satisfy the requirement that, for each number, it is either greater than all the numbers before it, or less than all the numbers before it. How many such permutations are there?

Problem 5.10 Consider the number 100000.

(a) How many factors does it have?

(b) How many ways are there to represent it as the product of 2 factors if we consider products that differ in the order of factors to be different?

(c) How many ways are there to represent it as the product of 3 factors if we consider products that differ in the order of factors to be different?

Problem 5.11 Suppose you have 8 red cards and 25 black cards. Assume all the cards of the same color are identical. Deal the cards out in a line.

(a) How many different arrangements of the cards are there?

(b) Repeat part (a), if there must be at least 2 black cards between all the red cards.

Problem 5.12 Suppose you place 9 rings on the 3 mid fingers of your left hand (that is, not on your thumb or your pinky). How many different outcomes are possible if

(a) all the rings are identical, and no finger has more than 3 rings?

(b) all the rings are different, and no finger has more than 3 rings?

(c) all the rings are identical, and no finger has more than 8 rings?

(d) all the rings are different, and all the rings are on a single finger?

Problem 5.13 How many ways are there to write 200 as the sum of three non-negative integers (we care about the order of the numbers)

(a) in total?

(b) if all three numbers must be different?

5.1.3 Binomial Coefficient Identities

Recall we can write the binomial coefficients in a form similar to Pascal's Triangle:

$$\binom{0}{0}$$
$$\binom{1}{0} \quad \binom{1}{1}$$
$$\binom{2}{0} \quad \binom{2}{1} \quad \binom{2}{2}$$
$$\binom{3}{0} \quad \binom{3}{1} \quad \binom{3}{2} \quad \binom{3}{3}$$
$$\binom{4}{0} \quad \binom{4}{1} \quad \binom{4}{2} \quad \binom{4}{3} \quad \binom{4}{4}$$
$$\binom{5}{0} \quad \binom{5}{1} \quad \binom{5}{2} \quad \binom{5}{3} \quad \binom{5}{4} \quad \binom{5}{5}$$

It will be useful to keep this relationship in mind for many of the identities below.

Symmetry

$$\binom{n}{k} = \binom{n}{n-k} \text{ for } 0 \le k \le n.$$

Pascal's Identity

$$\binom{n}{k} + \binom{n}{k+1} = \binom{n+1}{k+1}.$$

Reduction

$$k\binom{n}{k} = n\binom{n-1}{k-1} \text{ for } 0 \le k \le n.$$

Hockey Stick Identity

For n, r positive integers with $n > r$,

$$\binom{r}{r} + \binom{r+1}{r} + \binom{r+2}{r} + \cdots + \binom{n}{r} = \binom{n+1}{r+1},$$

or, using summation notation, $\displaystyle\sum_{k=r}^{n} \binom{k}{r} = \binom{n+1}{r+1}.$

Example 5.5

Suppose Tom has 8 friends, and he will invite some of them over for dinner (some can include none or all of his friends).

(a) How many different collections of friends could Tom invite over for dinner?

Solution: $2^8 = \binom{8}{0} + \binom{8}{1} + \cdots + \binom{8}{8} = 256.$

He can either invite each friend or not. We can also use the Binomial Theorem.

(b) Suppose he invites either 5 or 6 friends. How many different collections are possible?

Solution: $\binom{8}{5} + \binom{8}{6} = 84 = \binom{9}{6}.$

We examine two cases, and can use Pascal's Identity.

Example 5.6

Suppose you have one friend named Gwen, as well as 5 other friends. You want to invite a group of 3 friends out to dinner. How many different groups of friends could you invite with or without Gwen?

Solution: $\binom{5}{2} + \binom{5}{3} = \binom{6}{3}.$

Consider cases based on whether Gwen is invited or not. The same argument works with n other friends and inviting a group of k to dinner to prove Pascal's Identity.

Example 5.7

Consider a city grid below:

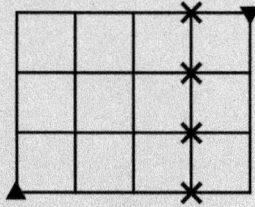

How many paths are there from the lower left triangle to the upper right triangle?

Solution: $\binom{3}{3} + \binom{4}{3} + \binom{5}{3} + \binom{6}{3} = 1 + 4 + 10 + 20 = 35$.

There are $\binom{7}{4}$ such paths. We can break these paths into cases based on which $\times$ they travel through *last* during the path. (If a specific $\times$ is the last travelled through, the path must have gone right from that $\times$, so there is only one way to complete the path to the upper right triangle.)

The same argument can be expanded to show the Hockey Stick Identity.

Binomial Theorem

Let n be a positive integer, then

$$(a+b)^n = \binom{n}{0}a^n + \binom{n}{1}a^{n-1}b + \cdots + \binom{n}{k}a^{n-k}b^k + \cdots + \binom{n}{n-1}ab^{n-1} + \binom{n}{n}b^n,$$

or, using summation notation, $(a+b)^n = \displaystyle\sum_{k=0}^{n} \binom{n}{k}a^{n-k}b^k$..

Example 5.8

Find the coefficient of x^3y^2 in $(x+y)^5$.

Solution: $\binom{5}{3} = 10$.

Problem 5.14 What is the coefficient of x^2 in $(x+3)^8$?

Problem 5.15 Practice on Pascal's Triangles.

(a) Show that $\dbinom{8}{0} + \dbinom{8}{2} + \dbinom{8}{4} + \dbinom{8}{6} + \dbinom{8}{8} = \dbinom{8}{1} + \dbinom{8}{3} + \dbinom{8}{5} + \dbinom{8}{7}$.

(b) Compute $2\dbinom{8}{0} + 2\dbinom{8}{2} + 2\dbinom{8}{4} + 2\dbinom{8}{6} + 2\dbinom{8}{8}$.

Problem 5.16 Find the constant term in the expansion of $\left(\sqrt{x} + \dfrac{1}{\sqrt{x}} - 2\right)^5$.

Problem 5.17 Simplify the following

(a) $\displaystyle\sum_{k=1}^{n} k\dbinom{n}{k}$.

(b) $\displaystyle\sum_{k=0}^{n} \frac{1}{k+1}\dbinom{n}{k}$.

5.2 **Further Practice on Counting**

Problem 5.18 Suppose 8 dinner guests attend a dinner party and are seated at a circular table. Two of the guests are a couple that must sit together. If one of the seats is the "Head of the Table", how many ways are there to seat the guests? Hint: Be careful!

Problem 5.19 How many numbers between $1 - 100$ are a multiple of either $5, 7, 9,$ or 11?

Problem 5.20 Suppose you want to bring a collection of 12 sodas to a party. You can choose from 6 types (and all that matters is how many of each soda you bring). How many ways can you do this if

(a) there are no restrictions on the types of soda you bring?

(b) one of your friends at the party really likes Fanta (one of the 6 types) so you want to bring at least 6 Fantas to the party?

Problem 5.21 Suppose you have 12 identical balls and 4 numbered boxes. How many ways are there to put the balls in the boxes if the first box has at least 3 balls, but no more than 5 balls?

Problem 5.22 Suppose you have a sequence of 10 integers, and the first integer is 10. Every integer after the first is either one larger or smaller than the previous.

(a) How many different sequences of integers are possible?

(b) How many different possibilities are there for the last integer in the sequence?

Problem 5.23 In how many ways can a necklace be made using 5 identical red beads and 2 identical blue beads? Hint: Try brute force.

Problem 5.24 Suppose you have 20 people and need to form 3 (non-empty) committees. The number of people on each committee must be a multiple of 5. If all three committees are different (Executive, Party Planning, etc.), how many ways can the committees be chosen?

Problem 5.25 Suppose 10 girls and 20 boys sit around a table. The girls are grouped in 5 pairs (which sit together), and in between each pair is at least 2 boys. How many arrangements are there? Hint: Work slowly, and break the problem into pieces.

Problem 5.26 Suppose you have 8 different books. You want to use the books as gifts for 3 of you friends. How many ways are there to give out the gifts? Be nice: each friend gets at least one book! (Every gift is determined only by which books are given.)

Problem 5.27 A train with 15 passengers must make 15 stops.

  (a) How many ways are there for the passengers to get off the train at the stops, if not all the passengers get off at the same stop?

(b) Repeat part (a) if we only care about the number of passengers getting off at each stop?

Problem 5.28 Suppose 5 people get in an elevator on Floor 0. The people leave the elevator somewhere between (inclusive) Floors 1 and Floor 10.

(a) If we only care about how many people get of at each floor, how many ways can the people get off?

(b) Suppose the 5 people all get off at different floors. If we only care about what collection of floors the elevator stops on, how many different collections are there?

Problem 5.29 Suppose we have four black, four white, and four green balls. How many ways are there to put the 12 balls into 6 distinguishable boxes if every color is put in at least 2 boxes?

Problem 5.30 How many different ways are there to represent 81000 as the product of 3 factors if we consider products that differ in the order of factors to be different?

Problem 5.31 Suppose an ant starts at the origin $(0,0)$. Ever step it takes is either $(1,1)$ or $(1,-1)$ (so it moves diagonally up and diagonally down).

(a) How many different ways can the ant move from the origin to $(20,0)$?

(b) Repeat part (a) if the ant makes a stop at $(10,0)$ along the way.

Problem 5.32 Suppose you have 8 numbered red cards and 20 identical black cards. How many ways are there to arrange the cards so that there is at least two black cards between each red card.

Problem 5.33 Suppose you have 6 numbered red cards and 20 numbered black cards. How many ways are there to arrange the cards in a circle if there must be at least 2 black cards between each red card?

Problem 5.34 Suppose you have the numbers $\{0,1,2,3,4,5\}$. How many 6-digit numbers can be formed with 1 next to 0 or 2? (Use each number exactly once.) Caution: 012345 is *not* a 6 digit number so be careful!

Problem 5.35 Suppose 4 people Albert, Bill, Charles, and Drew run a race. It was predicted that Albert would finish first, Bill second, and Charles third. How many outcomes of the race are there where all 3 of these predictions are wrong? (Note: We are *not* making any predictions about Drew.)

Problem 5.36 Suppose you give out 5 distinct books to 3 of your friends. Each friend gets at least one book. How many ways can you give out the books? Do the calculation using

(a) PIE.

(b) directly using cases.

Problem 5.37 Suppose a pizza place has 10 toppings available. You want to order 2 different 3-topping pizzas. Suppose repeated toppings are not allowed on a single pizza, and the order of the toppings on a pizza does not matter. If you only care which two pizzas you get, how many ways are there to make the order?

Problem 5.38 How many ways are there to put 8 balls in 4 numbered boxes, so that each box gets at least one ball? Hint: It is probably easiest to use PIE and Complementary Counting.

Problem 5.39 Suppose you place 4 different rings on the 3 mid fingers of your left hand. How many different outcomes are possible? Hint: Cases!

5.3 Probability Concepts

5.3.1 Definitions

- Suppose we have an experiment whose outcome depends on chance. The outcome of the experiment is called a *random variable*. The *sample space* of the experiment is the set of all possible outcomes. If the sample space is either finite or countably infinite (such as the positive integers), the random variable is said to be *discrete*.
- Generally a sample space is denoted by the capital Greek letter Ω. The sample space Ω corresponds to the set of possible outcomes of the experiment. A random variable is usually represented by a capital letter, such as X.
- The elements of a sample space are called *outcomes*. Each subset of a sample space is defined to be an *event*.
- Normally, the outcomes are denoted by lower case letters and events by capital letters.
- Two events A and B are called *disjoint* (or *mutually exclusive*) if $A \cap B = \emptyset$.
- The *probability* of an event A , denoted $\Pr(A)$, in the sample space Ω is a function defined on the subsets of Ω and satisfying the following:
  - $0 \le \Pr(A) \le 1$ for $A \subset \Omega$.
  - $\Pr(\Omega) = 1$.
  - If A and B are two disjoint events, $\Pr(A \cup B) = \Pr(A) + \Pr(B)$. This can be extended to infinitely many disjoint events.

  The *complement* of an event A, denoted $\overline{A}$ or A^c, represents the event where A does not happen. The probability $\Pr(\overline{A}) = 1 - \Pr(A)$.

Example 5.9

A die is rolled once. Let X denote the outcome of this experiment. Then the sample space for this experiment is the set

$$\Omega = \{1, 2, 3, 4, 5, 6\},$$

where each outcome i, for $i = 1, \ldots, 6$, corresponds to the number of dots on the face which turns up. The event

$$E = \{1, 3, 5\}$$

corresponds to the statement that the result of the roll is an odd number. The event E can also be described by saying that X is odd. Unless there is reason to believe the die is loaded, it is assumed that every outcome is equally likely. This means each of the six outcomes has a probability of $1/6$. The probability of the event E above is: $\Pr(E) = 1/2$.

5.3.2 **Classic Model and Geometric Model**

Classic Model. Suppose the sample space Ω is finite. Let A be an event, that is, a subset of Ω. The probability of A is

$$\Pr(A) = \frac{\text{want}}{\text{total}} = \frac{n(A)}{n(\Omega)},$$

where $n(A), n(\Omega)$ are the sizes of A, Ω respectively.

Example 5.10

Suppose you flip a fair coin 8 times.
  (a) What is the probability you get 5 heads?
      Solution: $\binom{8}{5}/2^8 = \frac{56}{256} = \frac{7}{32}$.
  (b) What is the probability you get an equal number of heads and tails?
      Solution: $\binom{8}{4}/2^8 = \frac{70}{256} = \frac{35}{128}$.

Geometric Model. If the values of the random variable is continuous, we sometimes can use geometry to represent the experiment and calculate the probabilities. Suppose you have a geometric shape Ω (in $1, 2, 3$ dimensions) and a subset A of this shape (still in the same number of dimensions). If you randomly pick one point in Ω, then the probability that this point belongs to A is

$$\Pr(A) = \frac{\text{want}}{\text{total}} = \frac{\text{size}(A)}{\text{size}(\Omega)},$$

where the "size" of A is its length, area, or volume depending on dimensions.

Example 5.11

Suppose a circle is inscribed in a square. If you randomly pick a point inside the square, what is the probability it is within the circle?
Solution: $\frac{\pi}{4} \approx 0.785$.
If the square has side length 2, the circle has radius 1, we then do area of circle (π) divided by area of square (4).

Example 5.12

Two people arrive at a party independently at random times between 6pm and 7pm and each stay for m minutes. What is m if there is a 40% chance that they meet at the party?

Solution:

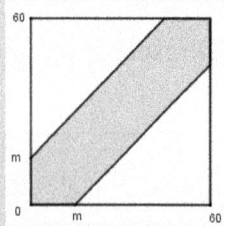

The x-axis give the arrival time of one (in minutes after 6pm), the y-axis the arrival time of the other. The arrival times can differ by up to m minutes, so the shaded band gives the arrival pairs (x,y) for which they meet. Thus we need $(60-m)^2/60^2 = 0.6$, giving $m^2 - 120m + 1440 = 0$. solve for m, we get $m = 60 \pm \sqrt{2160}$. But $m < 60$, so $m = 60 - 12\sqrt{15}$.

5.3.3 Principle of Inclusion-Exclusion

Let A, B, C be events, then

$$
\begin{aligned}
\Pr(A \cup B) &= \Pr(A) + \Pr(B) - \Pr(A \cap B); \\
\Pr(A \cup B \cup C) &= \Pr(A) + \Pr(B) + \Pr(C) \\
&\quad - \Pr(A \cap B) - \Pr(B \cap C) - \Pr(C \cap A) \\
&\quad + \Pr(A \cap B \cap C).
\end{aligned}
$$

The Principle of Inclusion-Exclusion can be extended to any number of events.

Problem 5.40 Let $\Omega = \{a, b, c\}$ be a sample space, where $\Pr(a) = \dfrac{1}{2}, \Pr(b) = \dfrac{1}{3}, \Pr(c) = \dfrac{1}{6}$. Find the probabilities of all eight possible events.

Problem 5.41 A die is loaded in such a way that the probability of each face turning up is proportional to the number of dots on that face. (For example, the probability of getting a six is twice that of getting a three.) What is the probability of getting an odd number in one throw?

Problem 5.42 Let A and B be events where $\Pr(A) = 1/2$, $\Pr(\overline{B}) = 1/3$, and $\Pr(A \cap B) = 1/4$. Find the value of $\Pr(A \cup B)$.

Problem 5.43 Two fair dice are rolled once, and the results are added. What is the probability that the sum is a prime number?

Problem 5.44 A reader of Marilyn vos Savant's column wrote in with the following question:

> My dad heard this story on the radio. At Duke University, two students had received A's in chemistry all semester. But on the night before the final exam, they were partying in another state and didn't get back to Duke until it was over. Their excuse to the professor was that they had a flat tire, and they asked if they could take a make-up test. The professor agreed, wrote out a test and sent the two to separate rooms to take it. The first question (on one side of the paper) was worth 5 points, and they answered it easily. Then they flipped the paper over and found the second question, worth 95 points: 'Which tire was it?' What was the probability that both students would say the same thing? My dad and I think it's 1 in 16. Is that right?

Is the reader's answer correct? If yes, explain. If no, give the right answer.

5.3.4 **Independence**

Two events A and B are said to be independent if $\Pr(A \cap B) = \Pr(A) \cdot \Pr(B)$.

Usually, independence means the event A does not physically influence event B.

Example 5.13

Two men, A and B are shooting a target. The probability that A hits the target is $\Pr(A) = \dfrac{1}{4}$, and the probability that B shoots the target is $\Pr(B) = \dfrac{1}{5}$, one independently of the other. Find
  (1) That A misses the target.
  (2) That both men hit the target.
  (3) That at least one of them hits the target.
  (4) That none of them hits the target.
Solution:
  (1) $\Pr(\overline{A}) = 1 - \dfrac{1}{4} = \dfrac{3}{4}.$
  (2) $\Pr(A \cap B) = \Pr(A) \cdot \Pr(B) = \dfrac{1}{4} \cdot \dfrac{1}{5} = \dfrac{1}{20}.$
  (3) $\Pr(A \cup B) = \Pr(A) + \Pr(B) - \Pr(A \cap B) = \dfrac{1}{4} + \dfrac{1}{5} - \dfrac{1}{20} = \dfrac{2}{5}.$
  (4) $\Pr(\overline{A} \cap \overline{B}) = \Pr(\overline{A \cup B}) = 1 - \Pr(A \cup B) = 1 - \dfrac{2}{5} = \dfrac{3}{5}.$

Problem 5.45 Events A and B are independent, events A and C are disjoint, and events B and C are independent. If $\Pr(A) = \dfrac{1}{2}, \Pr(B) = \dfrac{1}{4}, \Pr(C) = \dfrac{1}{8}$, find $\Pr(A \cup B \cup C)$.

Problem 5.46 Two numbers X and Y are chosen at random, and with replacement, from the set $\{1,2,3,4,5,6,7,8,9\}$. Find the probability that $X^2 - Y^2$ is an even number.

Problem 5.47 Three dice are rolled once. Find the probability of getting at least one six.

5.3.5 Conditional Probability

Given two events A and B.

The probability that event A happens given that event B has occurred is denoted as $\Pr(A|B)$, and defined as

$$\Pr(A|B) = \frac{\Pr(A \cap B)}{\Pr(B)}.$$

Note: If two events A and B are independent, then $\Pr(A|B) = \Pr(A)$.

The following identities are often useful:

$$\Pr(A \cap B) = \Pr(B)\Pr(A|B) = \Pr(A)\Pr(B|A).$$

Example 5.14

Ten cards numbered 1 through 10 are shuffled and then one card is chosen at random. If this card is an even numbered card, what is the probability that its number is divisible by 3?

Solution: Let event B be "the card is even numbered", and event A be "the card's number is divisible by 3". Then $A \cap B$ is the event "the card's number is divisible by 6". So the desired conditional probability is $\Pr(A|B) = \dfrac{1}{5}$.

Example 5.15

In a census, one family is randomly selected and there are two children in the family. Assume that it is equally likely for a child to be a boy or a girl. Given that at least one of the two children is a girl. What is the probability that both children are girls?

Solution: There are totally four outcomes in the sample space: $\{BB, BG, GB, GG\}$. The event "at least one of the children is a girl" is the subset $\{BG, GB, GG\}$. The event "both children are girls" is the subset $\{GG\}$. Therefore the conditional probability is $\dfrac{1}{3}$.

5.3.6 Total Probability and The Bayes Formula

Assume that events $B_1, B_2, \ldots, B_n$ are pairwise disjoint, and $B_1 \cup B_2 \cup \cdots \cup B_n = \Omega$. For any event A, the probability can be expressed in the following *Total Probability Formula*:

$$\Pr(A) = \Pr(A|B_1)\Pr(B_1) + \Pr(A|B_2)\Pr(B_2) + \cdots + \Pr(A|B_n)\Pr(B_n).$$

On the other hand, given that event A has occurred, we can go back and calculate the conditional probability for each B_i using the following *Bayes Formula*:

$$\Pr(B_i|A) = \frac{\Pr(A|B_i)\Pr(B_i)}{\Pr(A)} = \frac{\Pr(A|B_i)\Pr(B_i)}{\sum_{k=1}^{n}\Pr(A|B_k)\Pr(B_k)}.$$

Example 5.16

(2012 AMC 10B/12B) Suppose that one of every 500 people in a certain population has a particular disease, which displays no symptoms. A blood test is available for screening for this disease. For a person who has this disease, the test always turns out positive. For a person who does not have the disease, however, there is a 2% false positive rate—in other words, for such people, 98% of the time the test will turn out negative, but 2% of the time the test will turn out positive and will incorrectly indicate that the person has the disease. Let p be the probability that a person who is chosen at random from this population and gets a positive test result actually has the disease. Find p.

Solution: Let B be the event that a person has the disease, so from the given data, $\Pr(B) = \frac{1}{500}$. Therefore $\Pr(\overline{B}) = \frac{499}{500}$. Let A be the event that the test result is positive. Then we are told that $\Pr(A|B) = 1$, and $\Pr(A|\overline{B}) = 2\%$. The problem asks for $p = \Pr(B|A)$.

We first use the Total Probability Formula to find out $\Pr(A)$.

$$\Pr(A) = \Pr(A|B)\Pr(B) + \Pr(A|\overline{B})\Pr(\overline{B}) = 1 \cdot \frac{1}{500} + 0.02 \cdot \frac{499}{500} = \frac{549}{25000}.$$

Then we use Bayes Formula to find out $\Pr(B|A)$.

$$p = \Pr(B|A) = \frac{\Pr(A|B)\Pr(B)}{\Pr(A)} = \frac{1/500}{549/25000} = \frac{50}{549}$$

which is what we wanted.

Problem 5.48 Let A and B be two events and $\Pr(A) = 0.6$, $\Pr(B) = 0.7$. What are the maximum and minimum possible values of $\Pr(A \cap B)$?

Problem 5.49 Given ten cards with numbers $1, 2, \ldots, 10$ on them, each with one number. Randomly select 3 of the ten cards. What is the probability that ...

(a) the smallest number is 5?

(b) the largest number is 5?

Problem 5.50 There are 17 marbles in a bag. Among the marbles, 10 are white, 4 are black, and 3 are red. Nine marbles are taken out at random. What is the probability that 4 white, 3 black, and 2 red marbles are taken out?

Problem 5.51 Given 5 distinct pairs of shoes, and randomly select 4 shoes among them. What is the probability that at least one pair of shoes are chosen?

Problem 5.52 Put the 11 letters of "Probability" on 11 cards, and draw 7 cards successively. Find the probability that the 7 cards, in the order drawn, form the word "ability".

Problem 5.53 Randomly put 3 distinct balls into 4 distinct cups. What is the probability that ...

(a) the largest number of balls in each cup is 1.

(b) the largest number of balls in each cup is 2.

(c) the largest number of balls in each cup is 3.

Problem 5.54 Let A and B be events, and $\Pr(\overline{A}) = 0.3, \Pr(B) = 0.4, \Pr(A \cap \overline{B}) = 0.5$, find $\Pr(B|(A \cup \overline{B}))$.

Problem 5.55 Two dice are rolled once. Given that the sum of the two values is 7, determine the probability that one of the dice shows one.

Problem 5.56 Based on statistics, certain family (Dad, Mom, and one Child) has the following probabilities for catching some contagious disease: the probability that the Child is sick is 0.6; the probability that Mom is sick given that Child is sick is 0.5; the probability that Dad gets sick given that Mom and Child are both sick is 0.4. Determine the probability that Mom and Child are both sick but Dad is not.

Problem 5.57 Ten coins are in a bag, two of which are counterfeit. Choose one coin, without putting back, then choose another. What is the probability of each of the following events:

(a) Both are authentic.

(b) Both are counterfeit.

(c) One authentic, and one counterfeit.

(d) The second coin is counterfeit.

Problem 5.58 You want to call a friend, but forgot the last digit of the phone number. You decide to try dialing the last digit randomly.

   (a) What is the probability that you succeed in at most 3 tries?

   (b) Suppose you know that the last digit is an odd number. What is the probability that you succeed in at most 3 tries?

Problem 5.59 Two bags are given: the first bag contains n white balls and m red balls; the second bag contains N white balls and M red balls. Take one ball from the first bag and put into the second bag, then take one ball from the second bag. What is the probability that this ball is white?

Problem 5.60 Two boxes are given, where the first box contains 5 red balls and 4 white balls, and the second box contains 4 red balls and 5 white balls. Take 2 balls from the first box to put into the second box, then take one ball from the second box. What is the probability that this ball is white?

Problem 5.61 The trade mark of a certain product is "MAXAM", which is mounted on the wall. However, two of the letters fell off and janitor picked up the two letters and randomly put them back on. What is the probability that the sign still says "MAXAM"?

Problem 5.62 Suppose that 5% of men are color blind, and 0.25% of women are color blind. From a crowd of equal number of men and women, select one person randomly, and this person happens to be color blind. What is the probability that this is a man?

Problem 5.63 A student takes the AMC 12A and 12B. If he passes at least one of these tests, he will qualify for AIME. Assume the probability that he passes the AMC 12A is p. If he passes the AMC 12A, then the probability that he also passes the AMC 12B is also p. If he fails the AMC 12A, because of the stress, his chance of passing the AMC 12B is only $p/2$.

(a) What is the probability that he is qualified for AIME?

(b) Given that he has passed the AMC 12B. What is the probability that he passed the AMC 12A?

Problem 5.64 (Monty Hall Problem, based on Marilyn vos Savant's column) Suppose you're on a game show, and you're given the choice of three doors: Behind one door is a car; behind the others, goats. You want to choose the car. You pick a door, say No. 1 (but the door is not opened), and the host, who knows what's behind the doors, opens another door, say No. 3, which has a goat. He then says to you, "Do you want to pick door No. 2?" Is it to your advantage to switch your choice?

Problem 5.65 Given a line segment of length 1, find two points X and Y on it to split the segment into three pieces. What is the probability that the middle piece is shorter than 1/3?

Problem 5.66 Suppose a bag contains 7 red, 6 green, and 5 yellow balls. Suppose you pick 5 balls at once (so in no particular order).

(a) What is the probability you get 3 red, 1 green, and 1 yellow ball?

(b) What is the probability you get 0 yellow balls?

(c) What is the probability you get all 5 balls of the same color?

(d) What is the probability you do not get all red and green balls?

(e) What is the probability you get at least 2 red, and at least 1 green and at least 1 yellow ball? Hint: Be careful here!

Problem 5.67 Suppose you flip a coin 8 times. What is the probability you get more heads than tails?

Problem 5.68 Suppose you are dealt a five card hand from a standard deck of 52 cards (with 13 ranks $2, 3, \cdots, 10, J, Q, K, A$ and 4 suits: hearts, diamonds, clubs, and spades). Find the probability of:

(a) a flush (all cards of the same suit).

(b) a straight (all 5 ranks in a row, so $A, 2, 3, 4, 5$ up to $10, J, Q, K, A$).

(c) a full house (3 cards of one rank, 2 cards of another).

(d) a four of a kind.

(e) exactly two pairs. (That is, not four of a kind or a full house.)

(f) exactly one pair. (That is, not two pair or three of a kind, etc.)

Problem 5.69 Suppose you randomly pack a cooler (in no particular order) of 40 sodas (chosen from Coke, Sprite, and Fanta). Find the probability that:

(a) there are 20 Cokes and 20 Sprites.

(b) there are equal amounts of all sodas.

(c) there are at least 10 Cokes.

(d) there are at least 5 of each soda.

Problem 5.70 Suppose Jack and Jill both randomly come to school between 10 AM and 2 PM.

(a) What is the probability Jack comes to school before 11 AM?

(b) What is the probability Jack and Jill both come to school before 11 AM?

(c) What is the probability Jack comes to school before Jill?

(d) What is the probability Jack and Jill come to school within 30 minutes of each other?

Problem 5.71 Suppose Jackie is at a track meet. She is an erratic long jumper: each of her jumps are randomly between 2 and 8 meters.

(a) What is the probability a jump is more than 7 meters?

(b) What is the probability the jump is a integer length when measured in centimeters?

(c) If Jackie has two tries to jump more than 5 meters, what is the probability she succeeds (at least once)?

5.4 Distribution and Expected Values

Definitions

- A *probability distribution* is a function that describes the probability of a random variable taking certain values.
  For most of this book we will discuss discrete random variables.

Example 5.17

Throwing a fair die, each of the six values 1 to 6 has the probability 1/6, thus the distribution is $\Pr(X = 1) = \Pr(X = 2) = \cdots = \Pr(X = 6) = 1/6$. ♣

- For a discrete random variable X, $\sum_a \Pr(X = a) = 1$. The sum is over all possible values of the random variable X.

Example 5.18

Roll a die twice. Find the distribution of the minimum of the two values.
Solution: The possible results for the minimum of the two values are $\{1,2,3,4,5,6\}$. Calculating the probability for each value, we get $\Pr(X = 1) = 11/36, \Pr(X = 2) = 1/4, \Pr(X = 3) = 7/36, \Pr(X = 4) = 5/36, \Pr(X = 5) = 1/12, \Pr(X = 6) = 1/36$. ♣

- The *mode* is the most frequently occurring value in a distribution.
- The *expected value* of a random variable is the weighted average of all possible values this random variable can take on.
  Suppose random variable X can take value a_1 with probability p_1, value a_2 with probability p_2, and so on, up to value a_k with probability p_k. Then the expected value of this random variable X is defined as

$$E[X] = a_1 p_1 + a_2 p_2 + \cdots + a_k p_k.$$

Example 5.19

Roll a die twice. The expected value of the minimum of the two values is

$$1 \times \frac{11}{36} + 2 \times \frac{1}{4} + 3 \times \frac{7}{36} + 4 \times \frac{5}{36} + 5 \times \frac{1}{12} + 6 \times \frac{1}{36} = \frac{91}{36}.$$

Problem 5.72 There are 5 balls in a bag, numbered 1,2,3,4,5. Take 3 balls from the bag, and let X represent the largest number on the 3 balls. What is the distribution of X?

Problem 5.73 Roll a die twice. Find the distribution of the sum of the two values.

Problem 5.74 There are 15 balls in a bag, two of which are red and the remaining are white. Take 3 balls out of the bag. Let X be the number of red balls taken. Find the distribution of X.

Problem 5.75 Keith gets off work everyday at 5pm. He can choose to take the bus or subway to get home. From his experience, the probabilities of getting home at certain time range taking either subway or bus is in the following table:

Arrival time	5:35-5:39	5:40-5:44	5:45-5:49	5:50-5:54	5:55 & later
Subway	0.10	0.25	0.45	0.15	0.05
Bus	0.30	0.35	0.20	0.10	0.05

Today, he decided to flip a coin to choose between bus and subway, and then he arrived home at 5:47. What is the probability that he took the subway?

Problem 5.76 You are playing darts. Suppose the probability that you hit the target is 45%. Let X be the number of throws by the time you first hit the target. Find (1) the distribution of X, and (2) the probability that X is even.

Problem 5.77 A room has 3 windows of the same shape and size, only one of which is open. A bird flies into the room through the open window. It can only fly out through the open window. It flies around in the room and tries to find the open window. Assume that this bird does not remember which windows have been tried, and has the same probability to try any of the windows.

(a) Let X be the number of times the bird tries to fly through a window and finally gets out. Find the distribution of X.

(b) The owner said he has a bird that can memorize the windows it has tried, thus it tries each window at most once. Assume the owner is telling the truth, let Y be the number of tries this smart bird uses to get out the room. Find the distribution of Y.

(c) Calculate $\Pr(X < Y)$.

(d) Calculate $\Pr(Y < X)$.

Problem 5.78 Adam and Bob played darts. Adam's probability of hitting target is 0.6, and Bob's is 0.7. They each throw 3 times. Find the probability that ...

(a) they hit the same number of times.

(b) Adam hits more than Bob.

Problem 5.79 Normally, we assume that a new born child has a 50-50 chance to be a boy or a girl. Suppose, in a certain country, because of the desires for each family to have sons, a policy was made by the government: Any family with a son already is not allowed to have more children. Any family with all girls is allowed to have one more child, and so on. (1) So roughly among the children, are there more boys or more girls? (2) Find the expected number of children in each family.

Problem 5.80 Each of three friends decides to go to a fitness club at least once in a 5-day work week. Each of them select a day at random. What is the probability that they go in 3 separate days?

Problem 5.81 Three hunters shot at a rabbit. If the probabilities of hitting the target was 0.6, 0.5, and 0.4 for each of them respectively, what is the probability that the rabbit was hit?

Problem 5.82 Only one key is needed to open the door, and David has five keys, two of which can open the door. The problem is that he doesn't know which two. David tries the five keys one by one. What is the probability that he opens the door within three tries?

Problem 5.83 A fair coin is flipped repeatedly, until the first heads occurs.

 (a) What is the probability exactly two flips are needed? Three flips? Four flips?

(b) What is the probability that exactly n flips are needed?

Problem 5.84 Same situation as the previous question. What is the expected value of number of flips until heads occurs?

Problem 5.85 Frank and Ed play 5 games during a contest. If either of them wins 3 games, the contest is over. If for each game, the probability that Frank wins is $\frac{2}{3}$, and the probability that Ed wins is $\frac{1}{3}$. What is the probability that Frank wins the contest $3 : 1$?

Problem 5.86 Two points are randomly selected on a circle. What is the probability that the chord connecting the two points is longer than the radius?

Problem 5.87 There are 5 distinct red balls and 4 distinct white balls in a bag. Ed takes a ball from the bag and then places it back. He does this three times. What is the probability that he gets a red ball twice and a white ball once?

Problem 5.88 Randomly put n distinct balls into n boxes. Find the probability of getting exactly one empty box.

Problem 5.89 Randomly put 3 letters into 3 envelopes that are already addressed. Assume that each letter corresponds to only one correct address. Find the probability that at least one letter is in the correct envelope.

Problem 5.90 Ten people want to go to a basketball game, but there is only one ticket. So the ticket is put in one of 10 envelopes that look exactly the same, and the ten people pick up the envelopes one by one. Is it better to pick up an envelope earlier or later?

Problem 5.91 A fair coin is flipped repeatedly until two heads in a row appears. What is the probability that exactly 4 flips are needed?

Problem 5.92 Same situation as the previous question. What is the expected value of number of flips until two heads in a row occurs?

Answer Keys to Problems

The answer keys to the exercise problems in all the chapters are provided here. Only short answers are included. For problems requiring proofs, the solutions are omitted. Full solutions for all problems, including calculation problems and proof problems, can be found in the solution manual: *"Cracking the High School Math Competitions Solution Manual"*, by the same authors of this book.

Chapter 2. Number Theory Answer Keys

Problem 2.1 $2^6 - 63 = 1$.

Problem 2.2   (a) 1350^3

(b) 27000.

(c) $27000n^3$ for all positive integers n.

Problem 2.3 The perfect squares.

Problem 2.4 84 years

Problem 2.5   (a) 31726.

(b) 212.

(c) 11259375.

(d) 114.

Problem 2.6 7.

Problem 2.7 12,10,6,6.

Problem 2.8 31 Oct = 25 Dec, interpreted as "31 in octal (base 8) equals 25 in decimal (base 10)".

Problem 2.9 $37.

Problem 2.10 503118, and 503811.

Problem 2.11 12345678900954. Answer is not unique.

Problem 2.12 3.

Problem 2.13 9876504

Problem 2.14 301246

Problem 2.15 No.

Problem 2.16 2.

Problem 2.17 8880.

Problem 2.18 Use Euclidean algorithm.

Problem 2.19 8181.

Problem 2.20 7,43,271,331.

Problem 2.21 Use Bezout's identity: There exist integers a,b so that $\gcd(m,n) = am+bn$.

Problem 2.22 $p = 3, q = 2$.

Problem 2.23 123654, 321654.

Problem 2.24 120^{2793}.

Problem 2.25 List all possible remainders k for each modulus, and calculate the remainders of k^2 or k^3 and find out the possible results.

Problem 2.26   (a) 0,1

  (b) 0,1,4,7

Problem 2.27 7.

Problem 2.28 0.

Problem 2.29 All of them are 3 (mod 4), but squares are all 0 or 1 mod 4.

Problem 2.30 No.

Problem 2.31 No.

Problem 2.32 4.

Problem 2.33 Consider the numbers in mod 9. All of them are 1 mod 9. If one is a multiple of another, it would be 2 or 3 or 4 or 5 or 6 or 7 times the other, not possible to remain 1 mod 9.

Problem 2.34 3.

Problem 2.35 9.

Problem 2.36 5.

Problem 2.37 249.

Problem 2.38 25 and 76.

Problem 2.39 (20,21,29); (29,420,421).

Problem 2.40 (8,15,17); (15,112,113); (15,20,25); (9,12,15); (15,36, 39).

Problem 2.41 Yes, 2027 and 2029 are prime.

Problem 2.42 $A = B = 4$.

Problem 2.43 701239

Problem 2.44 153846.

Problem 2.45 11111111100.

Problem 2.46 5.

Problem 2.47 63,65.

Problem 2.48 6.

Problem 2.49 453420,413424, 373428.

Problem 2.50 987652413.

Problem 2.51 1,6,8,10,14,15,21,22,26,27,33,34,35,38,39,46.

Problem 2.52 5,12,14,15,16,17,18,19.

Problem 2.53 100.

Problem 2.54 4.

Problem 2.55 56.

Problem 2.56 3.

Problem 2.57 6.

Problem 2.58 7.

Problem 2.59 $a = 1, b = 8$.

Problem 2.60 89.

Problem 2.61 $a = 3, b = 2$.

Problem 2.62 27.

Problem 2.63   (a) 5.
  (b) None.

Problem 2.64 7.

Problem 2.65 36.

Problem 2.66 0.

Problem 2.67 $(115, 552), (232, 435)$.

Problem 2.68 $(1, 1), (3, 3)$.

Problem 2.69 16.

Problem 2.70 All positive even numbers.

Problem 2.71 (C).

Problem 2.72 $3 \mid n$.

Problem 2.73 4.

Problem 2.74 7.

Problem 2.75 13 and 20.

Problem 2.76 Infinitely many.

Problem 2.77 27.

Problem 2.78 The first statement is always true.

Problem 2.79 (B).

Problem 2.80 (C).

Problem 2.81 $(2, 59, 2)$, and $(11, 2, 23)$.

Problem 2.82 6.

Problem 2.83 17.

Problem 2.84 2 or 3.

Problem 2.85 2, 1, 8.

Problem 2.86 358

Problem 2.87 56.

Problem 2.88 $12005 = 5 \times 7^4$

Problem 2.89 10

Problem 2.90 2.

Problem 2.91 100 and 101.

Problem 2.92 1,2,5,10,20,25,50, 100, 125, 200, 250, 500, 1000, 1250.

Problem 2.93 3, 5, 9.

Problem 2.94 Yes for 99^{99}, no for 99!.

Problem 2.95 $2, 3, \ldots, 30, 31$.

Problem 2.96 Assume possible, then $80a = b + c$. Contradiction.

Problem 2.97 Assume all odd, and remove the first two and last two digits, and repeat until only one digit left.

Problem 2.98 The sum of the numbers of dimes and pennies is always odd. So the difference between them is also odd.

Problem 2.99 No.

Problem 2.100 315000000.

Problem 2.101 pair up k and $9999 - k$.

Problem 2.102 Reverse order and pair up.

Problem 2.103 $n = 3$, and the sets are $\{1/2, 2/3, 6/7, 41/42\}$, $\{1/2, 2/3, 7/8, 23/24\}$, $\{1/2, 2/3, 8/9, 17/18\}$, $\{1/2, 2/3, 9/10, 14/15\}$, $\{1/2, 3/4, 4/5, 19/20\}$, and $\{1/2, 3/4, 5/6, 11/12\}$.

Problem 2.104 434

Problem 2.105 No.

Problem 2.106 No.

Problem 2.107 Odd.

Problem 2.108 Yes.

Problem 2.109 No.

Problem 2.110 1987.

Problem 2.111 1324, 1423, 2314, 2413, 3412.

Problem 2.112 432

Problem 2.113 The 72nd card.

Problem 2.114 38

Problem 2.115 61.

Problem 2.116 2, 5, 7

Problem 2.117 $p = 3, q = 2$

Problem 2.118 $a = 256, b = 128$

Problem 2.119 406

Problem 2.120 259980.

Problem 2.121 No.

Problem 2.122 No.

Problem 2.123 $\dfrac{7}{332} = \dfrac{1}{48} + \dfrac{1}{3984}$.

Problem 2.124 35964.

Problem 2.125 9504

Problem 2.126 9.

Problem 2.127 No.

Problem 2.128 33

Problem 2.129 Proof omitted.

Problem 2.130 Odd.

Problem 2.131 101.

Chapter 3. Algebra Answer Keys

Problem 3.1 $5xy^2(2x - 3y + 5z)$.

Problem 3.2 $6x(a - b)^4(a - b + 5)$.

Problem 3.3   (a) $-2x^{n-1}y^n(x^n - y)^2(x^n + y)^2$
  (b) $(x - 2y - z)(x^2 + 4y^2 + z^2 + 2xy + xz - 2yz)$
  (c) $= (a - b + c)^2$
  (d) $= (a + b)^2(a - b)(a^4 - a^3b + a^2b^2 - ab^3 + b^4)$

Problem 3.4 $(a^{16} + b^{16})(a^8 + b^8)(a^4 + b^4)(a^2 + b^2)(a + b)(a - b)$.

Problem 3.5 $(x^8 + 1)(x^4 + 1)(x^2 + 1)(x + 1)$.

Problem 3.6   (a) $(2x - 1)(x - 3)$.
  (b) $(3k + 1)(k - 2)$.
  (c) $(2p + 3)(p - 1)$.

Problem 3.7   (a) $(x - 1)(x^2 + x + 1)(x^6 + 2x^3 + 3)$.
  (b) $(mn + m - n + 1)(mn - m + n + 1)$.
  (c) $(3x^2 + 1)(x^2 + 3)$
  (d) $(a^2 - ab + 1)(b^2 + ab + 1)$.

Problem 3.8 $k = 1985$.

Problem 3.9 0.

Problem 3.10 $m < 4$.

Problem 3.11 $x^2 - \dfrac{7}{6}x + \dfrac{1}{3} = 0$.

Problem 3.12 $-5 + \sqrt{30}$ and $-5 - \sqrt{30}$.

Problem 3.13 Proof omitted.

Problem 3.14 $p < -1$.

Problem 3.15 $m = 3$.

Problem 3.16 $k = 3$.

Problem 3.17 $\sqrt{5} - 2$.

Problem 3.18 0 or 16.

Problem 3.19   (a) $-2 < m \le -1$
(b) $m < -2$.

Problem 3.20 $-\sqrt{2}/2 < k \le 1$.

Problem 3.21 $x = 1, y = 1, z = 0$.

Problem 3.22 1.

Problem 3.23 ± 1

Problem 3.24 6.

Problem 3.25 $p^2 + 4q + q^2$

Problem 3.26 $-1/2$

Problem 3.27 $-5 < m \le -4$.

Problem 3.28 -5

Problem 3.29 $x = 1/2$ and $x = -1$.

Problem 3.30 $x = -6$ and $x = 5$.

Problem 3.31 $x = 7$ and $x = -5$.

Problem 3.32 $x = 2$ and $x = 4/3$.

Problem 3.33 $x = 0$ and $x = -3$.

Problem 3.34 $x = 1$, $x = 2$, and $x = 1/2$.

Problem 3.35 $-4, -2, 0, 2$.

Problem 3.36 $x = \pm\sqrt{6}$.

Problem 3.37 $x = 4$ and $x = -1$.

Problem 3.38 $x = 3$ and $x = -6$.

Problem 3.39 $4, -1$.

Problem 3.40 414

Problem 3.41 18.

Problem 3.42 5.

Problem 3.43 $k > -1/16$ and $k \neq 0$.

Problem 3.44 $a = 1, b = -1/2$.

Problem 3.45 $a = 0$ or $a > 25/4$.

Problem 3.46 $b = -28$.

Problem 3.47   (a)
   (b) $m = 0$ or $m = \dfrac{17}{12}$.

Problem 3.48 Yes, $m = -2$.

Problem 3.49 $\dfrac{13}{-2\sqrt{3}} = -\dfrac{13\sqrt{3}}{6}$.

Problem 3.50 $\sqrt{2} + \sqrt{3}$.

Problem 3.51 $\dfrac{15}{7}$.

Problem 3.52 $\dfrac{5}{3}$.

Problem 3.53 $x_{\max} = \sqrt[3]{2}/2$.

Problem 3.54 $x = \sqrt{3} - 1$ and $x = \pm\sqrt[4]{3}$.

Problem 3.55 $\dfrac{\sqrt{17} + 1}{2}$.

Problem 3.56 $\dfrac{\sqrt{21} - 1}{2}$.

Problem 3.57 $(2x^2 + 5x + 12)(2x + 7)(x - 1)$

Problem 3.58 $(x + 2)(x + 4)(x^2 + 5x + 8)$

Problem 3.59 $(2x + 1)(x - 2)(3x - 1)(x + 3)$.

Problem 3.60 $(x^2 - xy + y^2)^2$

Problem 3.61 $n(n + 1)(n + 2)(n + 3) + 1 = (n^2 + 3n + 1)^2$.

Problem 3.62 $(x + 2)(x + 3)(x - 5)$.

Problem 3.63 Proof omitted.

Problem 3.64 1.

Problem 3.65 $(x^2 + 3)(x^2 + 5x - 3)$.

Problem 3.66 $(x^2 - 6x + 17)(x^2 + 6x + 19)$.

Problem 3.67 $(x - 1)(x + 2)^2$

Problem 3.68 $(x^2 - 3xy - y^2)(x^2 + 3xy - y^2)$

Problem 3.69 $(x + 2)(x + 3)(x + 4)$

Problem 3.70 $x(x - 3)(2x + 3)(2x - 3)$.

Problem 3.71 $(x^2 + 3x + 1)(x^2 + 4x + 1)$.

Problem 3.72 $(x^2 + 4x - 7)(x^2 + 4x + 5)$

Problem 3.73 $(2x-3y+4)(x+3y+5)$

Problem 3.74 $(a^2+a+1)^2$.

Problem 3.75 $(a+2b)(d+2c)(x+2y)$.

Problem 3.76 $(ac+bd)(bc-ad)$.

Problem 3.77 $(a^2+a+1)(a^3-a^2+1)$.

Problem 3.78 $(x+y+z)(x-y+z)(x+y-z)(x-y-z)$.

Problem 3.79 $(a+1)^2(a^2+a+1)^2(a^2-a+1)^2$.

Problem 3.80 $(x+2)(x+6)(x^2+8x+10)$

Problem 3.81 1996.

Problem 3.82 (a) A nonzero constant.
   (b) **0**.
   (c) A value for the variable where the polynomial's value equals 0. In other words, a root of the polynomial equation.

Problem 3.83 (a) -7
   (b) 7
   (c) 0
   (d) 0.
   (e) 9.
   (f) 90000.
   (g) -14.

Problem 3.84 $7\cdot 2=14$.

Problem 3.85 $\sqrt{5}$.

Problem 3.86 $m=\dfrac{5-\sqrt{17}}{2}$.

Problem 3.87 365.

Problem 3.88 $1/2$.

Problem 3.89 $m = 0, n = -1$ or $m = 1, n = -3$.

Problem 3.90 $97 - 56\sqrt{3}$.

Problem 3.91 -23.

Problem 3.92 1.

Problem 3.93 6.

Problem 3.94 $4 \times 3 \times 1$.

Problem 3.95 $-(-2)/(-1990) = -1/995$.

Problem 3.96 -2.

Problem 3.97 -17.

Problem 3.98 $4 - \sqrt{13}$.

Problem 3.99 $(3,4), (4,3), (-2+\sqrt{3}, -2-\sqrt{3}), (-2-\sqrt{3}, -2+\sqrt{3})$.

Problem 3.100 96.

Problem 3.101 $16 \times 8 = 128$.

Problem 3.102 -41.

Problem 3.103 5.

Problem 3.104 22.

Problem 3.105 36.

Chapter 4. Geometry Answer Keys

Problem 4.1   (a) Rectangle. Many examples are possible.
(b) Rhombus. Many examples are possible.

Problem 4.2 Proof omitted.

Problem 4.3 Proof omitted.

Problem 4.4   (a) Yes.
(b) $90°$.

Problem 4.5 $15°$.

Problem 4.6 $75°$

Problem 4.7 $90°$.

Problem 4.8 Yes, it is true.

Problem 4.9 No.

Problem 4.10 $90°$.

Problem 4.11 (B)

Problem 4.12 It is true in all cases.

Problem 4.13 No.

Problem 4.14 23.

Problem 4.15 $24\sqrt{3} - 36$.

Problem 4.16 $43\sqrt{3}/4$.

Problem 4.17 $12(\sqrt{2} - 1)$.

Problem 4.18 $36°, 72°, 72°$.

Problem 4.19 600.

Problem 4.20 $5(\sqrt{3}+1)$.

Problem 4.21 Yes.

Problem 4.22 $\dfrac{29}{2}$.

Problem 4.23 Proof omitted.

Problem 4.24 Proof omitted.

Problem 4.25 $\dfrac{120}{13}$.

Problem 4.26 54.

Problem 4.27 Proof omitted.

Problem 4.28 Proof omitted.

Problem 4.29 $1/2$.

Problem 4.30 $1/8$.

Problem 4.31 Proof omitted.

Problem 4.32 $18\sqrt{2}$.

Problem 4.33 $2/3$.

Problem 4.34 100.

Problem 4.35 Proof omitted.

Problem 4.36 π.

Problem 4.37 $36/5$.

Problem 4.38 $2 \times 5, 2 \times 6$ or $1 \times 10, 1 \times 12$.

Problem 4.39 $2\sqrt{s(s-a)(s-b)(s-d)}$, where $s = (a+b+d)/2$.

Problem 4.40 Proof omitted.

Problem 4.41   (a) Proof omitted.
  (b) $1/2$.

Problem 4.42   (a) Proof omitted.
  (b) Proof omitted.

Problem 4.43 Proof omitted.

Problem 4.44 Proof omitted.

Problem 4.45 Proof omitted.

Problem 4.46 Proof omitted.

Problem 4.47   (a) Proof omitted.
  (b) Proof omitted.
  (c) Proof omitted.
  (d) Proof omitted.

Problem 4.48 The fixed value is the height from B to $\overline{AC}$.

Problem 4.49 18.

Problem 4.50 $\dfrac{27+9\sqrt{3}}{\pi^2}$.

Problem 4.51 $\dfrac{3-\sqrt{3}}{3}$.

Problem 4.52 Proof omitted.

Problem 4.53 Proof omitted.

Problem 4.54 84.

Problem 4.55 Proof omitted.

Problem 4.56 $1/16$.

Problem 4.57 $1:10$.

Problem 4.58 $\pi/6$.

Problem 4.59 12.

Problem 4.60 $6 \cdot \dfrac{1}{3} = 2$.

Problem 4.61 $14\sqrt{5}$.

Problem 4.62 Proof omitted.

Problem 4.63 Proof omitted.

Problem 4.64   (a) 4.
  (b) $3 + 2\sqrt{2}$.

Problem 4.65   (a) $4\pi/3$.
  (b) $2\pi/3 - \sqrt{3}/2$.

Problem 4.66   (a) $\sqrt{2} - 1$.
  (b) $1/3$.

Problem 4.67   (a) 5.
  (b) $100 - 25\pi$.

Problem 4.68 Proof omitted.

Problem 4.69 $5 + 2\sqrt{3} + 2\sqrt{2}$.

Problem 4.70   (a) 2π.
  (b) $\pi - \sqrt{3}/2$.

Problem 4.71   (a) $2\sqrt{3} - 3$.
  (b) $1/2$.

Problem 4.72   (a) Proof omitted.

(b) Proof omitted.

Problem 4.73 Proof omitted.

Problem 4.74 $90°$.

Problem 4.75 $2\sqrt{10}$.

Problem 4.76 $18 + 9\sqrt{2}$.

Problem 4.77 $81\sqrt{3} - 81\pi/2$.

Problem 4.78 Proof omitted.

Problem 4.79 Proof omitted.

Problem 4.80 $40°$.

Problem 4.81 (a) $2/\sqrt{3}$.
(b) $\frac{2}{3}\sqrt{21}$.

Problem 4.82 $120°$.

Problem 4.83 (a) Proof omitted.
(b) Proof omitted.

Problem 4.84 Proof omitted.

Problem 4.85 $8/13$.

Problem 4.86 $8\sqrt{10}/5$.

Problem 4.87 Proof omitted.

Problem 4.88 Proof omitted.

Problem 4.89 $3, 12$.

Problem 4.90 8.

Problem 4.91   (a) $\dfrac{1}{6}\pi$.

(b) $\dfrac{\sqrt{3}}{2}\pi$.

Problem 4.92 32.

Problem 4.93   (a) Proof omitted.
(b) Proof omitted.

Problem 4.94   (a) $6, 9, 5$.
(b) $7 : 8$ for both.

Problem 4.95   (a) $2\sqrt{33}$.
(b) $4\sqrt{33}/5$.

Problem 4.96 $4\sqrt{2} + 4\sqrt{3}$.

Problem 4.97   (a) $\sqrt{2}$.
(b) $1/2$.

Problem 4.98   (a) $\dfrac{4\sqrt{3}}{27}\pi$.

(b) $\dfrac{\sqrt{3}}{9}$.

Problem 4.99 144.

Problem 4.100   (a) 4.
(b) $1/6$ or $1/3$.

Problem 4.101   (a) Two square faces (both different), 8 triangular faces (2 groups of 4 congruent triangles); 8 vertices; 16 edges.
(b) 20.

Problem 4.102   (a) $19, 54$.
(b) $56, 84, 30$.

Problem 4.103   (a) 6 vertices, 12 edges, and 8 faces.
(b) $2/3$.

(c) 1 or $\sqrt{3}/3$.

Problem 4.104   (a) $2\sqrt{17}$.
(b) $\sqrt{17}/2$.

Problem 4.105   (a) 8.
(b) $18, 14$.

Problem 4.106   (a) 3.
(b) $3/2$.

Problem 4.107 Proof omitted.

Problem 4.108 10.

Problem 4.109   (a) Proof omitted.
(b) $\sqrt{rR}$.

Problem 4.110 $1 : 3 : 3 : 5$.

Problem 4.111 2.

Problem 4.112   (a) $-\sqrt{2}/2$
(b) $-1/2$
(c) 1
(d) $1 + \sqrt{2}$
(e) $(\sqrt{2} + \sqrt{6})/4$

Problem 4.113   (a) $\cos x$
(b) $-\sin x$

Problem 4.114   (a) $\dfrac{\sqrt{2}}{2}$
(b) $24/25$
(c) $\sqrt{2}/2$
(d) $(\sqrt{2} + \sqrt{6})/4$
(e) $\sqrt{3}$.
(f) $-1/16$.
(g) 2.
(h) $\pi/4$.

(i) $3/2$.

Problem 4.115 $\cos(A)$.

Problem 4.116 1.

Problem 4.117 $\sin^2 x$.

Problem 4.118 $1/5$.

Problem 4.119 $-\dfrac{1}{2}$.

Problem 4.120 $1 + \sqrt{5}/3$.

Problem 4.121 $\dfrac{\sqrt{3}+1}{2}$

Problem 4.122 2

Problem 4.123 $\dfrac{63}{65}$.

Problem 4.124 (a) $\pm\sqrt{3}/2$.
  (b) ± 5.
  (c) $\pm\sqrt{a^2 + b^2}$.
  (d) $\pm\sqrt{2 + \sqrt{2}}$.

Problem 4.125 1.

Problem 4.126 $(\tan x)^{\cot x} < (\tan x)^{\tan x} < (\cot x)^{\tan x} < (\cot x)^{\cot x}$.

Problem 4.127 $-\dfrac{117}{125}$.

Problem 4.128 $\dfrac{16}{65}$.

Problem 4.129 $\pi/2$.

Problem 4.130 (a) 1.

(b) $\cos 2x$.

(c) $\dfrac{\cos \dfrac{(n+1)x}{2} \sin \dfrac{nx}{2}}{\sin \dfrac{x}{2}}$.

(d) $\arctan(n+1)$.

(e) $\arctan(2n+1) - \dfrac{\pi}{4}$, or equivalently, $\arctan \dfrac{n}{n+1}$.

Problem 4.131 $\dfrac{\sqrt{3}}{2}$.

Problem 4.132 $(\pi/3, \pi/2) \cup (3\pi/4, 3\pi/2) \cup (5\pi/3, 7\pi/4)$

Problem 4.133 (a) $\pi/6 + k\pi, k \in \mathbb{Z}$.
(b) $\pi/2 + k\pi, 2k\pi \pm \pi/3, k \in \mathbb{Z}$.
(c) $\pi/3, 2\pi/3, 4\pi/3, 5\pi/3$
(d) $k\pi - \arctan 3, k\pi + \pi/4, k \in \mathbb{Z}$.

Chapter 5. Combinatorics Answer Keys

Problem 5.1 76.

Problem 5.2   (a) $5 \cdot 8^6 = 1310720$.

(b) $5 \cdot 7 \cdot 6 \cdot 5 \cdot 4 \cdot 3 \cdot 2 = 5 \cdot \dfrac{7!}{1!} = 25200$.

(c) $5 \cdot 8^5 \cdot 4 = 655360$.

(d) $5 \cdot 6 \cdot 5 \cdot 4 \cdot 3 \cdot 2 \cdot 2 + 4 \cdot 6 \cdot 5 \cdot 4 \cdot 3 \cdot 2 \cdot 2 = 5 \cdot \dfrac{6!}{1!} \cdot 2 + 4 \cdot \dfrac{6!}{1!} \cdot 2 = 12960$.

Problem 5.3   (a) $12! = 479001600$.

(b) $\dfrac{12!}{4!} = \dbinom{12}{4} \cdot 8! = 19958400$.

(c) $3! \cdot 10! = 21772800$.

(d) $12! - 3! \cdot 10! = 457228800$.

Problem 5.4   (a) $\dbinom{15}{5} \cdot \dbinom{10}{5} = 756756$.

(b) $\left(\dbinom{25}{5} - \dbinom{15}{5} - \dbinom{10}{5} \right)^2 = 2487515625$.

(c) $\left(\dbinom{25}{5} - \dbinom{15}{5} - \dbinom{10}{5} \right) \cdot \dbinom{20}{5} = 773262000$.

Problem 5.5   (a) $\dbinom{6}{0} + \dbinom{5}{1} + \dbinom{4}{2} + \dbinom{3}{3} = 1 + 5 + 6 + 1 = 13$.

(b) 32.

Problem 5.6   (a) $20! = 2432902008176640000$.

(b) $(5! \cdot 7! \cdot 8!) \cdot 3! = 146313216000$.

(c) $15! \cdot (16 \cdot 15 \cdot 14 \cdot 13 \cdot 12) = 15! \cdot \dfrac{16!}{11!} = 685430596730880000$.

Problem 5.7   (a) $\dbinom{40 + 8 - 1}{40} = 62891499$.

(b) $\dbinom{24 + 8 - 1}{24} = 2629575$.

(c) 1.

(d) $\dbinom{30 + 7 - 1}{30} = 1947792$.

Problem 5.8 (a) $20! \cdot \dfrac{21!}{11!}$.

(b) $\dbinom{20}{10} \cdot \dbinom{21}{10}$,

(c) $\dfrac{20!}{20} \cdot \dfrac{20!}{10!}$.

Problem 5.9 2^{n-1}.

Problem 5.10 (a) 36.

(b) 36.

(c) 441.

Problem 5.11 (a) $\dbinom{33}{8} = 13884156$.

(b) $\dbinom{11+9-1}{11} = 75582$.

Problem 5.12 (a) 1.

(b) $9!$.

(c) $\dbinom{9+3-1}{9} - 3 = 52$.

(d) $3 \cdot 9! = 1088640$.

Problem 5.13 (a) $\dbinom{200+3-1}{200} = 20301$.

(b) $\dbinom{200+3-1}{200} - 303 = 19998$.

Problem 5.14 $\dbinom{8}{6} \cdot 3^6 = 20412$.

Problem 5.15 (a)

(b) $2^8 = 256$.

Problem 5.16 $(-1)^5 \dbinom{10}{5} = -252$.

Problem 5.17 (a) $n2^{n-1}$.

(b) $\dfrac{2^{n+1}-1}{n+1}$.

Problem 5.18 $2! \cdot \dfrac{7!}{7} \cdot 8 = 11520.$

Problem 5.19 $20 + 14 + 11 + 9 - 2 - 2 - 1 - 1 - 1 - 1 = 46.$

Problem 5.20 (a) $\dbinom{12 + 6 - 1}{12} = 6188.$

(b) $\dbinom{6 + 6 - 1}{6} = 462.$

Problem 5.21 $\dbinom{9 + 3 - 1}{9} + \dbinom{8 + 3 - 1}{8} + \dbinom{7 + 3 - 1}{7} = 136.$

Problem 5.22 (a) $2^9.$
(b) 11.

Problem 5.23 3.

Problem 5.24 $3 \cdot \dfrac{20!}{10! \cdot (5!)^2} = 3 \cdot \dbinom{20}{10} \cdot \dbinom{10}{5} \cdot \dbinom{5}{5} = 139675536.$

Problem 5.25 $(2!)^5 \cdot \dfrac{5!}{5} \cdot \dbinom{10 + 5 - 1}{10} \cdot 20! = 1870337211021939179520000.$

Problem 5.26 $3^8 - 3 \cdot 2^8 + 3 = 5796.$

Problem 5.27 (a) $15^{15} - 15 = 437893890380859360.$
(b) $\dbinom{15 + 15 - 1}{15} - 15 = 77558745.$

Problem 5.28 (a) $\dbinom{5 + 10 - 1}{5} = 2002.$

(b) $\dbinom{10}{5}.$

Problem 5.29 $\left(\dbinom{4 + 6 - 1}{4} - 1 \right)^3 = 2000376.$

Problem 5.30 $\dbinom{4 + 3 - 1}{4}^3 = 3375.$

Problem 5.31   (a) $\binom{20}{10} = 184756$.

(b) $\binom{10}{5}^2 = 63504$.

Problem 5.32 $8! \cdot \binom{6+9-1}{6} = 121080960$.

Problem 5.33 $\dfrac{6!}{6} \cdot \binom{8+6-1}{8} \cdot 20! = 375737386142800281600000$.

Problem 5.34 366.

Problem 5.35 11.

Problem 5.36   (a) $3^5 - 3*2^5 + 3 = 150$.

(b) $\binom{3}{2} \cdot \dfrac{5!}{1! \cdot 1! \cdot 3!} + \binom{3}{2} \cdot \dfrac{5!}{2! \cdot 2! \cdot 1!} = 150$.

Problem 5.37 $\left(\binom{\binom{10}{3}}{2} \right) = \binom{120}{2} = 7140$.

Problem 5.38 $4^8 - 4 \cdot 3^8 + 6 \cdot 2^8 - 4 = 40824$.

Problem 5.39 360.

Problem 5.40 $\Pr(\emptyset) = 0, \Pr(a) = \dfrac{1}{2}, \Pr(b) = \dfrac{1}{3}, \Pr(c) = \dfrac{1}{6}, \Pr(a,b) = \dfrac{5}{6}, \Pr(a,c) = \dfrac{2}{3},$ $\Pr(b,c) = \dfrac{1}{2}, \Pr(\Omega) = 1$.

Problem 5.41 3/7

Problem 5.42 11/12.

Problem 5.43 5/12

Problem 5.44 No. It should be 1/4.

Problem 5.45 $\dfrac{23}{32}$.

Problem 5.46 $\dfrac{41}{81}$

Problem 5.47 $1 - \left(1 - \dfrac{1}{6}\right)^3 = \dfrac{91}{216}$.

Problem 5.48 0.6 and 0.3.

Problem 5.49 (a) $\dbinom{5}{2} / \dbinom{10}{3} = 1/12$.

(b) $\dbinom{4}{2} / \dbinom{10}{3} = 1/20$.

Problem 5.50 $\dfrac{\dbinom{10}{4}\dbinom{4}{3}\dbinom{3}{2}}{\dbinom{17}{9}} = \dfrac{252}{2431}$.

Problem 5.51 $1 - \dbinom{5}{4} \cdot 2^4 / \dbinom{10}{4} = \dfrac{13}{21}$.

Problem 5.52 $\dfrac{1 \cdot 2 \cdot 2 \cdot 1 \cdot 1 \cdot 1 \cdot 1}{{}_{11}P_7}$.

Problem 5.53 (a) $\dfrac{{}_4P_3}{64} = \dfrac{3}{8}$.

(b) 9/16.
(c) 1/16.

Problem 5.54 0.25.

Problem 5.55 1/3.

Problem 5.56 0.18.

Problem 5.57 (a) 28/45
(b) 1/45
(c) 16/45
(d) 1/5

Problem 5.58   (a) $3/10$

   (b) $3/5$

Problem 5.59 $\dfrac{N+1}{M+N+1} \cdot \dfrac{n}{n+m} + \dfrac{N}{M+N+1} \cdot \dfrac{m}{n+m}.$

Problem 5.60 $53/99.$

Problem 5.61 $3/5.$

Problem 5.62 $20/21.$

Problem 5.63   (a) $\dfrac{3}{2}p - \dfrac{1}{2}p^2.$

   (b) $\dfrac{2p}{p+1}.$

Problem 5.64 It is. The probability of getting the car is increased from $1/3$ to $2/3$ if you switch.

Problem 5.65 $5/9$

Problem 5.66   (a) $\dbinom{7}{3}\dbinom{6}{1}\dbinom{5}{1} / \dbinom{18}{5}.$

   (b) $\dbinom{13}{5} / \dbinom{18}{5}.$

   (c) $\left[\dbinom{7}{5} + \dbinom{6}{5} + \dbinom{5}{5} \right] / \dbinom{18}{5}.$

   (d) $1 - \left[\dbinom{13}{5} / \dbinom{18}{5} \right].$

   (e) $\left[\dbinom{7}{3}\dbinom{6}{1}\dbinom{5}{1} + \dbinom{7}{2}\dbinom{6}{2}\dbinom{5}{1} + \dbinom{7}{2}\dbinom{6}{1}\dbinom{5}{2} \right] / \dbinom{18}{5}.$

Problem 5.67 $\dfrac{93}{256}.$

Problem 5.68   (a) $\dbinom{4}{1} \cdot \dbinom{13}{5} / \dbinom{52}{5}.$

   (b) $10 \cdot 4^5 / \dbinom{52}{5}.$

(c) $\binom{13}{1}\binom{12}{1}\binom{4}{3}\binom{4}{2}/\binom{52}{5}$.

(d) $\binom{13}{1}\binom{12}{1}\binom{4}{4}\binom{4}{1}/\binom{52}{5}$.

(e) $\binom{13}{2}\binom{11}{1}\binom{4}{2}^2\binom{4}{1}/\binom{52}{5}$.

(f) $\binom{13}{1}\binom{12}{3}\binom{4}{2}\binom{4}{1}^3/\binom{52}{5}$.

Problem 5.69   (a) $1/\binom{40+3-1}{40}$.

(b) 0.

(c) $\binom{30+3-1}{30}/\binom{40+3-1}{40}$.

(d) $\binom{25+3-1}{25}/\binom{40+3-1}{40}$.

Problem 5.70   (a) $\dfrac{1}{4}$.

(b) $\dfrac{1}{16}$.

(c) $\dfrac{1}{2}$.

(d) $\dfrac{3}{4}$.

Problem 5.71   (a) $\dfrac{1}{6}$.

(b) 0.

(c) $\dfrac{11}{36}$.

Problem 5.72 $\Pr(X=3)=1/10, \Pr(X=4)=3/10, \Pr(X=5)=3/5$.

Problem 5.73 The sum can be $2,3,4,\ldots,12$, and the probability for each value is:
$1/36, 2/36, 3/36, 4/36, 5/36, 6/36, 5/36, 4/36, 3/36, 2/36, 1/36$.

Problem 5.74 $X=0,1,2$, and the probabilities are $22/35, 12/35, 1/35$.

Problem 5.75 9/13.

Problem 5.76 (1) $\Pr(X = k) = 0.45 \times (0.55)^{k-1}, k = 1, 2, \ldots,$ (2) $\dfrac{11}{31}$.

Problem 5.77 (a) $\Pr(X = k) = (1/3) \cdot (2/3)^{k-1}$.
(b) 1/3 for each of $Y = 1, 2, 3$.
(c) 8/27
(d) 38/81

Problem 5.78 (a) 0.32076
(b) 0.243

Problem 5.79 (1) Roughly it's the same, (2) 2.

Problem 5.80 12/25

Problem 5.81 0.88 (or 88%)

Problem 5.82 0.9 (or 90%)

Problem 5.83 (a) $1/4, 1/8, 1/16$.
(b) $1/2^n$

Problem 5.84 2

Problem 5.85 $\dfrac{8}{27}$

Problem 5.86 2/3

Problem 5.87 100/243

Problem 5.88 $\dbinom{n}{2} \cdot n!/n^n$

Problem 5.89 2/3.

Problem 5.90 Neither. Everyone has 1/10 probability to get the ticket.

Problem 5.91 1/8

Problem 5.92 6